Performance Contracting: Expanding Horizons 2nd Edition

Performance Contracting: Expanding Horizons

2nd Edition

Shirley J. Hansen, Ph.D.

Library of Congress Cataloging-in-Publication Data

Hansen, Shirley J., 1928-
Performance contracting: expanding horizons / Shirley J. Hansen.--2nd ed.
p. cm.
Includes bibliographical references and index.
ISBN 0-88173-530-2 -- ISBN 0-88173-531-0 (e-book)
1. Energy industries. 2. Contracting out. I. Title.

HD9502.U52H365 2006
658.2--dc22

2006040903

Published by The Fairmont Press, Inc.
700 Indian Trail
Lilburn, GA 30047
tel: 770-925-9388; fax: 770-381-9865
http://www.fairmontpress.com

Distributed by Taylor & Francis Ltd.
6000 Broken Sound Parkway NW, Suite 300
Boca Raton, FL 33487, USA
E-mail: orders@crcpress.com

Distributed by Taylor & Francis Ltd.
23-25 Blades Court
Deodar Road
London SW15 2NU, UK
E-mail: uk.tandf@thomsonpublishingservices.co.uk

Printed in the United States of America
10 9 8 7 6 5 4 3 2 1

0-88173-530-2 (The Fairmont Press, Inc.)
0-8493-9380-9 (Taylor & Francis Ltd.)

Contents

SECTION I—PERFORMANCE CONTRACTING TODAY 1

Chapter

1 MAKING THE BUSINESS CASE FOR ENERGY EFFICIENCY 3
The Tough Energy Efficiency Sell 5
The Anatomy of a Customer 6
As Management Sees It 7

2 ENERGY OPPORTUNITIES: THE MOVING TARGET 15
Performance Contracting in Retrospect 17
How ESCOs Work 20
Financial Models 21
Pushing the Envelope 27

3 MONEY MATTERS 31
Cost-effectiveness 32
Cost of Delay 35
Cost Avoidance 38

4 PARTNER SELECTION: FROM BOTH SIDES OF THE FENCE .. 41
Preparation: From the Owner's Side 41
Developing an RFQ/RFP 45
Establishing Criteria 53
The Evaluation Process 56
Preparation: The ESCO's Side 61

5 AN ESCO'S GUIDE TO MEASUREMENT AND VERIFICATION—Dr. Stephen Roosa 65
What Drives the Need for Measurement and Verification 66
The Uses for Measurement and Verification 68
Measurement & Verification Protocols 69
Cost of Measurement & Verification 72
The Measurement & Verification Process 74

6. FINANCING ENERGY EFFICIENCY 87
Creating Bankable Projects 89
Project Management 93
The Owner's Perspective 97
A Primer on Financing 100

7 QUALITY CONTRACTS 105
Laying the Groundwork 106
Planning Agreement 107
Energy Services Agreement 109
Key contract Considerations 113
Section by Section 118
The Schedules 121
Negotiations 122
If I Were on "Their" Side of the Table 125

8 WHERE THE SAVINGS ARE: PROJECT MANAGEMENT 127
ESCO Management Strategies 129
The Project Manager 130
Management Strategies for the Owner 134
The Energy Manager 137

9 COMMUNICATIONS STRATEGIES 143
Effective Communications 145
Communicating Energy Needs 148
Getting the Word Out 154
Getting the Job Done 155

SECTION II—MANAGING RISKS 157

10 CUSTOMER RISKS AND MITIGATING STRATEGIES 161
The Risk Analysis Framework 161
Performance Contracting Options 163
Keeping Risks in Perspective 176
The Big Risk: Getting the Right ESCO 176

11 ESCO RISKS AND MANAGEMENT STRATEGIES 179
ESCO Risk Vulnerability 179

Operations and Maintenance Practices 192
Measurement and Verification 193
Project Implementation 193
Managing Risks through the Financial Structure 195
The ESCO Fee 198

12 THE IAQ/ENERGY EFFICIENCY INTERFACE 201
Finding Answers 202
Determining the Value of Increased Outside Air 205
Virtues of Ventilation 208
The Real IAQ/Energy Efficiency Relationship 208
The Mutual Goal of IAQ and Energy Efficiency 210

SECTION III—
PERFORMANCE CONTRACTING:
THE NEXT GENERATION 213

13 EXPANDING THE ESCO MODEL 217
Options that Expand the Model 217
A Performance Contract that Increases
Energy Consumption 219
The All-Purpose ESCO 221

14 PRODUCTIVE ESCO/VENDER RELATIONS—
MICHAEL GIBSON AND BRIAN TODD 227
Owner's Perspective for ESCO Services 228
Initial Barriers to Vendor and ESCO Teaming 229
Vendor Traditional Role 232
Rethinking Delivery Mechanisms Offered by Design Build 234
Choosing the Correct Vendor-ESCO Partner 239
Conclusion 240

15 USING THE WEB FOR ENERGY DATA ACQUISITION &
ANALYSIS—Paul Allen, Dave Green and Jim Lewis 241
Energy Information Systems 241
EIS Implementation Options 247
Viewing and Using the Data 250

Case Study—Retail Store Lighting 251
Case Study—Chiller Plant Optimization 254

SECTION IV—ESCOS GO GLOBAL 261

16 ARE YOU READY TO GO INTERNATIONAL?—Don Smith
Questions and Answers about Going Global 265

17 ASSESSING FOREIGN OPPORTUNITIES FOR ESCOS—Jim Hansen 273
The International Marketplace
Preliminary Assessment 274
Country Analysis 275
In-Country Support 281
Market Assessment 286
Test the "Facts" 288

18 INTERNATIONAL FINANCING—TOM DREESSEN 289
A Need for Commercially Viable Project Financing 290
Capacity Building within Local Financial Institutions 293
An International Energy Efficiency Financing Model 294
Market-Based Incentives 296

19 THE GLOBAL PERSPECTIVE—BOB DIXON 299
Why Globalize? 299
What Is an ESCO? 299
What Are an ESCO's "Product and Services"? 300

APPENDIX A Cost-effectiveness and Cost of Delay Applications 303
APPENDIX B. Examples of Evaluation Criteria and Procedures 307
APPENDIX C. Sample Planning Agreement 310

GLOSSARY OF TERMS 315

INDEX 321

Acknowledgements

In the years since I wrote *Performance Contracting for Energy and Environmental Systems,* the world of energy performance contracting has expanded and merged into the global economy. When The Fairmont Press asked that I provide a second edition of *Performance Contracting: Expanding Horizons,* I welcomed the challenge. It is an opportunity to address the constantly changing world in which energy service companies (ESCOs) ply their trade.

As the ESCO world expands, it takes on many new facets. This book, and the struggles performance contracting has had and continues to have today, would not have been possible without the devoted help of so many of our colleagues.

My sincere thanks, therefore, go to the contributing authors, who are introduced in each section where their work appears. My deep appreciation to each of them, as they took time from their very busy schedules to offer us some very valuable insights. As you will see, they are the "cream of the crop" and have given this book a much broader perspective.

To those we have worked with, such as the World Bank and the United Nations Development Program in Brazil and the Dutch government in Mongolia, and US Agency for International Development in many countries, thank you for expanding the scope of all our visions.

We have stated previously that we learn daily from our clients—from ESCOs to end users—we again offer our deep appreciation for the pleasure of working with you and learning from you. Our experiences with you in the US and 35 other countries around the world have provided the opportunity for us to gain greater knowledge about our industry and to continue our own growth.

A special thank you goes to Jeannie C. Weisman Douglas for her work on the first edition of this book. She has retired from the industry, but her inspiration and insights are still missed.

The contributing authors have brought new dimensions and exciting perspectives to the book. I am deeply indebted to Stephen A. Roosa, Brian

Todd, Michael Gibson, Paul Allen, Dave Green, Jim Lewis, Tom Dreessen, Jim Hansen and Bob Dixon. While the excellent chapters they have contributed speak for themselves, the introductions to sections where their chapters appear contain an introduction to their exceptional qualification.

Illustrations can help make a point, highlight an idea, or (gasp) break up page after page of tedious print. If you can find an illustrator who has work in the Smithsonian and major collections around the world, it's even better. Should take a lot of persuasion for someone of that caliber to do illustrations for a technical book, but the powers of a mother should never be underestimated. Once again, I am deeply indebted to our son, Stephen Hansen, for sharing his gift with us.

A very heartfelt acknowledgement goes to James C. Hansen, who not only shared some of his expertise in Chapter 17, but edited the entire book. His excellent assistance in this and all of my work at home and overseas makes it all possible.

Finally, to all of you, who continue to make our own horizons expand, we thank you.

Section I

Performance Contracting Today

The concept of performance contracting is based on the educated guess. Despite the huge environmental and economic benefits offered by energy efficiency through performance contracting, owners and potential energy service companies are often hesitant to enter into this no-man's-land. To overcome such hesitancy, this book is designed to expand on the education and lessen the guessing.

This section leads off with a totally new piece, which focuses on how we can make the business case for energy efficiency. Making the tough energy efficiency sell appears to be a problem around the world—certainly in all 35 countries where we have worked. As stated in the chapter, it is positioned first because making the sale is vital to making anything else happen. Without the revenues, the remainder of the book becomes an academic exercise.

While common threads remain in the fabric of performance contracting, the weave, color and texture are constantly changing. The first section of this book captures some of the threads of the first edition and even reaches back to the beginnings of performance contracting. Those threads, pulled together through more recent spinning, create the rich texture of what performance contracting is today.

In this second edition, "Performance Contracting Today" shifts its focus to more pragmatic aspects of performance contracting that energy service companies (ESCOs) and owners face today. In particular, we are indebted to Dr. Stephen A. Roosa for giving us a valuable view from the trenches of "An ESCO's Guide to Measurement and Verification." Dr. Roosa is an account executive for Energy Systems Group in Louisville, Kentucky, and draws on more than 25 years of experience in energy efficiency and performance contracting. He is a member of the Measurement and Verification Professionals Certification Board and a past president of the Association of Energy Engineers.

Thoughts on "Communication Strategies" have also been drawn out of previous works and given their own chapter to highlight an area of great weakness in the ESCO industry. Effective communications and information management are consistently skills in which ESCOs fall short around the world.

The other sections of the book provide insights on the inevitable risks encountered in an industry that offers guarantees, and explores some intriguing ways to expand the ESCO offering. While it might be incredibly tempting to jump to some of the later chapters, the reader is strongly encouraged to first address Section I. It is the basis for the "education" so critical to performance contracting and the means to move closer to making guessing an art.

Chapter 1

Making the Business Case for Energy Efficiency

There is a line in the business world, "Nothing happens until somebody sells something." Taking this thought even further, one sale does not make a successful business any more than a grain of sand makes a beach.

The customers are as equally dependent on successful businesses as their energy service providers are. The energy service company (ESCO) must be there to back the guarantee and provide service and guidance through the life of the project. In fact, the expected longevity of the performance contractor is a critical criteria in ESCO selection.

Any business model, including the performance contracting model developed in this book, requires effective marketing strategies and sales techniques. Since "selling something" enables us to take the actions suggested throughout the book, it seems a good place to start.

Energy efficiency has many obvious benefits. In addition to preserving our limited fossil fuels, it reduces the client's operational costs as well as the organization's emission of pollutants. Even better, it can all be done through self-funded work. Energy efficiency (EE) offers a truly unique opportunity to make money while improving the environment. Consider it, "Doing well while doing good."

Money now paying the utility for wasted energy can be redirected into new equipment, facility / process modifications, lower operating costs, and a more competitive position in the market place. Non-profit organizations can have more money for their critical missions.

Simply put, "self-funded" means the needed investment capital comes from money already in the client's budget. It is hard to overstate to the client the benefits of using money already in the budget that is being wasted, to invest in a more efficient operation and reduce pollution at the same time. Don't let anyone compare this investment to those, which require new money to be allocated in the budget!

A major obstacle that frequently hinders the client's energy efficiency efforts is the lack of initial capital to do the EE work. Performance contracting provides that capital and expertise; and allows the customer to access future savings NOW for the immediate benefit of the client's operation.

With all this in mind, the question that inevitably emerges is: *If energy efficiency/performance contracting is such a win-win-win proposition, why is it such a tough sell?*

In countries around the world—transitional economies, industrial or developing countries—people in the energy efficiency field have trouble making the sale. Whether an energy policy maker in government, an energy manager in a corporation, or sales people of an ESCO, every energy professional shares the same frustrations. The sale to businesses, especially industry, has been particularly difficult. Since business is the backbone of the economy, the ultimate source of new revenue and a major consumer of energy, it is a sale that must be made. Further, if you can sell performance contracting to business management, chances are you can sell to anybody.

Zealous engineers, who eagerly try to sell energy efficiency and performance contracting (PC) to top management, typically find themselves relegated to the boiler room. Once the technical professional sees the exciting energy saving potential in a facility or process, he or she can get so caught up in all the technical benefits that the bigger picture sometimes escapes them.

The horrible truth is that top management is not interested in ENERGY! They don't want to hear about gigajoules or British thermal units. In fact, that stuff really turns them off. Try talking "energy" to a CEO or CFO and you are almost guaranteed a one-way ticket to the catacombs where the heating and cooling systems are housed.

To communicate effectively with management, three approaches structured to respond to the businessman's point of view are suggested: (1) viewing energy efficiency and conservation as a very cost-effective delivery system for meeting environmental mandates and / or social responsibility—a way to make money while reducing emissions; (2) positioning energy savings as a percentage of the bottom line; and (3) providing an effective cost / benefit analysis procedure, which compares the net benefits of energy efficiency and conservation to increased production.

Finally, we need to remind ourselves and top management as forcefully as possible that EE can be a self-funding endeavor. CEOs and CFOs have a tendency to compare energy investments to other business investments and fail to appreciate that no new money is required to do this work. When CEOs and CFOs start talking IRRs, hurdle rates, and ROIs with a two-year ceiling, it's a cinch they are trying to fit the EE investment into their regular investment model. Then, it is time to remind them once again that the money needed for energy investments is already in the budget—and being spent on wasted energy.

THE TOUGH ENERGY EFFICIENCY SELL

A basic tenent of any EE sales message should be:

Energy efficiency is an investment; not an expense.

Moreover, energy efficiency is a *very sound* investment. In the ESCO industry, we talk rather glibly about 2-year paybacks, but it's hard to find another source that gives you a 50% interest rate. Imagine the stampede to a bank that offered a 50% interest rate! Yet, our potential clients walk right by the opportunity on a daily basis.

EE is an incredibly cost-effective way to cut operating costs; and, through performance contracting, those costs can be cut without any up-front capital expenditures. Reducing operating costs is good for the customer, the market, and the country's economy. Further, studies have shown that energy efficiency (EE) creates five times as many jobs per megawatt hour as does the creation of new power generation—and for about one-eighth the capital investment. At the same time, energy efficiency is also reducing pollution emissions. We hear so much about our environmental needs and global warming, but so very, very little about the most cost-effective way to meet such needs.

With so many benefits, people should be lined up at the door; eager to invest in energy efficiency. But ironically, convincing people they should reduce energy consumption and save money is hard work. Why is it such a struggle to make the case for energy efficiency? And what can we do about it?

Several years ago, a dear friend who was in top management of a major corporation, gave me some sage advice, fundamental to our problem. He said, "Shirley, you folks must learn to fish from the fish's point of view." Good idea, but first we must consider the fish we are trying to catch.

THE ANATOMY OF A CUSTOMER

We have two body parts in our customer that need unique and special attention. The same sell will not work for both. Facility people, including O&M personnel, need special consideration. Then, we need to consider how we can sell effectively at the management level.

THE FACILITY PERSPECTIVE

In the eyes of facility managers and O&M personnel, you, as an EE vendor are usually not part of the solution. YOU ARE A BIG PART OF THE PROBLEM. A major task for every seller, who has contact with the customer, is to turn that around.

Against a backdrop of management that doesn't care and facility people who have every reason to resent you, we begin to get a picture of why energy efficiency is a tough sell.

Before we get to selling anything, we must first get the "fish's" attention. To do so, we must talk *their* language and make the case in *their* terms. Once we have their attention, we must learn to bait the hook with a particularly juicy morsel. What is attractive to our facility managers or O&M personnel "fish?"

Fishing in the Facility Pond

"Catching" the facility people is critical for two reasons. They may not be the ones who say "Yes" to a deal, but their "No" can kill it. If you get the deal and you have not won over the people in the trenches, they can make your life miserable.

If, on the other hand, the deal is structured to get them something *they* believe they need, it's amazing what can happen. A program that meets some key O&M needs can make an incredible difference. In addition to tools or specific O&M measures, the often unvoiced "need" is simply for a bit of the recognition that most facility people hunger for. For example, as a 'thank you' from one energy service company (ESCO), the firm took the facility people from the more successful project sites to a football game. The future savings more than paid for the tickets! A certificate can be less expensive, but can often do the job just as effectively if it is awarded to plant engineers, building custodians, etc. before their boss or their board. That piece of paper, often posted on the boiler room wall, can go a long way toward cementing a partnership. Furthermore, these people talk to their friends in other organizations and the word spreads. Some of an organization's best selling takes place after the contract is signed!!

But we are talking about getting the sale, so we need to get to them up front. The first law of effective selling is: LISTEN. Look around, the best salesmen are good listeners. A key part of the sale is finding out what the O&M folks want and finding a way for them to get it. Contracts make an excellent vehicle for earmarking some of the savings for O&M needs.

For many years, many of us, trying to sell energy efficiency to the CEO, thought we were turned away due to the discomfort top management felt when the subject of "energy" was introduced. It certainly played a part, but in retrospect we now realize that other concerns, often more important concerns, were at play.

As Management Sees It

The horrible truth is that top management is just not interested in *ENERGY!* They don't want to hear about Btu or kWh. Finally, we have figured it out: CEOs and CFOs DO NOT BUY ENERGY; *they buy what it can do.* They buy lighting, running motors, and processing. It's only when the switch is flipped and nothing happens that management becomes aware of the critical role energy plays in its operation.

We simply cannot get them to worry about using something more efficiently, which is basically non-existent in their lives. Whether the blackouts are in Italy or California, in an incredibly short time, it's back to business as usual while costly downtimes seem to be forgotten.

To fish from the fish's point of view, we must first realize that top management is interested in delivering promised results—be it student achievement, patient care, or selling widgets. Of these, the hardest sell around the world seems to be industry. A good place for us to concentrate our thoughts at this point.

Such details as "energy" are just noise to management—a small irritating noise for someone else to deal with. This noise factor is part of a much bigger problem: management is "facility blind." Managers can walk the corridors, but they seldom see the facility itself until something goes wrong.

The problem is further exacerbated by the facility people themselves. When the operations and maintenance (O&M) budget is cut, facility people look for ways to stretch what they've got. From the management point of view, things still look good; so the logical conclusion is that it was a good place to cut the budget. The better the facility people do their jobs the more invisible they become. And the more frustrated.

Then, someone selling energy efficiency wanders in with the resources facility people covet and proceeds to tell the boss about ways it could be done better. Does it come as any a surprise that some facility managers resent those EE salesmen? If we were to walk in their shoes for a while, we'd become more sensitive to their needs and in a better position to get their attention.

BAITING MANAGEMENT'S HOOK

No matter how enticing the bait, it doesn't do any good unless you first bring it to the fish's attention. So rule #1: GET MANAGEMENT'S ATTENTION.

When we go after the really big fish, we must listen very carefully and address their concerns. Listen to the folks who sell to management regularly, they will tell you that management is interested in money, being competitive, the budget, money, reducing operating costs, money, environment, and *money.*

If we carefully analyze what they are telling us, "money" would seem to be the key. Since the money for wasted energy goes up in smoke, one way to get their attention is to literally burn money. Pile some dollars/euros (on a fireproof tray of course) on the desk and put a match to it. If that is not legal, burn another country's money. Then, remind them that THEIR money, which is being paid for wasted energy, is currently going up the smokestack. *NEVER TO BE RECOVERED.*

Or, throw money around on the floor—all around the room. If you make the denominations big enough, someone is going to go pick up a few bills. Then you can start talking about the money per square foot/meter that are just laying around in their facilities. Money that will disappear if they don't get busy and do something about it.

SETTING THE HOOK

Once we have their attention, we need to back it up with something substantive. Consider: (1) energy efficiency and conservation are very attractive ways to help the client **meet environmental mandates** and/or be socially responsible making money while reducing emissions; (2) the client can be provided a new perspective of energy savings as a **percentage of the bottom line**; or (3) we can provide them an effective **cost/benefit analysis** procedure, which compares the net benefits of energy efficiency to increased production.

Environmental Benefits

It goes without saying that a good marketing strategy is to study the market and the individual customer before you try to make the sale. Part of that research should check if there are environmental mandates, or if there is pressure to meet social responsibility. If either situation exists, demonstrating how they can make money through energy efficiency while improving the environment can be very attractive.

Consider giving them a worksheet, which will help them calculate the potential amount of reduced energy consumption* and what that will equate to in money saved *as well as* reduced emissions. When they see the results of their own calculations, they are more apt to accept the idea.

Even better, determine if there is a market for emission reduction credits. Many countries are offering such incentives, which provides a growing financial opportunity.

The Bottom Line

If we really are to put ourselves in management's shoes, we need to look at the impact on the bottom line. If energy savings get lost in budget mumbo-jumbo, management will not be able to see EE's net benefit.

Look at energy as a raw material used to produce a unit of product, or deliver a particular service. If 30 percent of a cement factory's raw material costs go for energy supplies—and if you can reduce that amount by 20 percent, you can show management the direct EE benefit of 6 percent in the bottom line. That's huge! (Even when the initial cost of the energy efficiency efforts are factored in, the benefits can still be pretty spectacular.)

Then, as all good fishermen know, you have to "reel them in." So ask them to tell you of another approach that will bring 6 percent to the bottom line. Work through the options with them. You've got a winner—and pretty soon they will realize it!

The third way to approach management follows from this bottom line thinking.

Making the Case for EE

Particularly in industrial settings, energy efficiency is a hard sell because an investment in production has a better fit with management's way of doing business. More production and shiny new equipment make their hearts sing.

Analyze the situation. Chances are management got to the top by focusing on production. This is often referred to as a boxes mentality (versus a services mentality). People resist changing procedures that have brought them success; so the inclination is to just pile the boxes higher.

Given the proclivity to stick with the tried and true, this option needs to be approached with caution. We need to avoid "selling energy" *per se* and focus on REDUCED OPERATING COSTS. To business managers, reduced

*Accepted conversion factors for reduced emissions per unit of fuel can be obtained from the state and/or federal energy and environment offices.

operating costs can lead to; a) more competitive pricing in the market place; and/or b) a bigger net profit for the company.

So the key is to show management ways to:

(1) cost-effectively reduce operating costs; and then
(2) compare the benefits from energy efficiency to the benefits from increased production.

Below is a methodology that can help you walk the client through an example using customer information.

Reducing Operating Costs

First, we need to establish what would be a reasonable amount of savings for a specified time that can be achieved in a given enterprise. Hopefully, some kind of scoping audit, or knowledge of the market and processes, will provide workable numbers with which you, and they, are comfortable. Later, we'll develop an investment figure needed to achieve those savings. Finally, develop a net benefit profile over, say, five years. It

WORKSHEET #1

XYZ SHOE MANUFACTURERS

Net Benefit Profile

Estimated energy savings	700,000 /yr
Projected investment to achieve these savings 1.5 million … prorated over 5 years.	300,000 /yr
Net benefit	400,000 /yr
Net benefit over 5 years (until investment is paid off)	2,000,000 (a)

will look something like Worksheet 1.

It may pay to note that after the fifth year (after the debt has been paid), the net benefit to the customer will be $700,000/yr.

Increasing Revenue to Yield Same Benefit

Now we need to work with the client to determine what it would take to get the same net benefit through production. Starting with the net energy savings over 5 years (the number on line (a) in Worksheet 1), we need to calculate what it would take to obtain a revenue stream from business income that would achieve the same results. In order to develop this example further, we need to make a couple of assumptions about the XYZ Shoe Manufacturers. (In actual practice, you would use the numbers supplied by the client.) For our purposes here, let's assume XYZ Shoe Manufacturers has: 1) a margin* of 20%; and 2) the manufacturer gets $40/pair of shoes sold. With these assumptions, the next step is to determine how many additional pairs of shoes would the manufacturer need to sell to equal the net energy savings?

First, we determine how big the revenue stream would have to be to yield a $2 million with a margin at 20%. In other words, 20% of what figure equals $2 million. The answer is: $10 million.

Next question for the manufacturer is: How many shoes would you have to sell to bring in $10 million? The answer is: 250,000 *additional* pairs of shoes over five years, or 50,000 *additional* pairs per year.

Now, we get to the big question: How big an investment would be required in production equipment, packaging, delivery, advertising, sales personnel, etc. to sell 50,000 additional pairs of shoes EVERY YEAR for the next five years? Chances are very good that it will add up to more than $1.5 million.

The other half of the worksheet for calculating equivalent additional revenue, should look something like Worksheet 2 shown on the following page.

After such an investment is made, there is still the question as to what happens if those additional 250,000 pairs of shoes are NOT sold?

While going over the worksheet with the customer, it pays to remind them that in the revenue option, management must depend on someone else (the buyer) to make it happen. Further, they have not improved their

---

*Depending on the customer, you may find it more effective to work with *net profit*, but the percentage figure will not be as big and the resulting numbers will be smaller.

WORKSHEET 2

XYZ SHOE MANUFACTURERS

Equivalent Additional Revenue

Calculating revenue needed:	
Bring forward (a)	2,000,000
Margin	20%
Revenue stream need to = (a) @ 20%	10,000,000 (b)
Number of shoes that must be sold to equal (b)	250,000 pairs
Prorated over five years	50,000 pairs
Investments required to increase production by 50,000 pairs/yr (equipment, packaging, deliver, advertising, personnel, etc.)	???????????? (c)

How does (c) compare with the needed EE investment in Worksheet 1?

position in the market through more cost-effective operations. In fact, they may have weakened it as they try to recover the new production investment costs.

For comparison, with the cost reduction option, management depends on internal resources to make it happen. For our illustration, they have also reduced operational costs by $400,000/yr., which could give them a major competitive advantage in the market place.

Finally, our biggest challenge may be convincing top management that EE can be a self-funding endeavor. We must be sensitive to how they perceive things. CEOs and CFOs have a tendency to compare energy investments to other business investments and fail to appreciate that no new money may be required to do energy efficiency work. The $1.5 million energy investment in the above example comes from money currently going to the utility to pay for wasted energy. So the money needed for energy investments is already in the budget—and being spent needlessly. *Wasted energy is the financing source for energy efficiency.*

It's important to once more underscore the fact that the investment is right there in avoided utility costs. This is money, which will otherwise go up the smokestack, creating more pollution every day that the energy efficiency measures are not in place. In contrast, the money to increase production is new money that must be added to the budget—typically increasing unit product costs on the street, weakening their competitive position.

The business case for energy efficiency can be made, and made in a language management will understand. Just be sure you get the fish's attention and have the right bait on the hook. Then, remember to wet the line where the fish are.

Chapter 2

Energy Opportunities: The Moving Target

Some of us can remember when "energy" was tied to words like kinetic. In 1973, however, oil coming to the US was embargoed and overnight our *energy* world changed. We became conscious of fossil fuels, kilowatts, retrofits, and *energy*. Suddenly we became aware of how we depend on energy, and other countries, to supply it.

Unfortunately, that awareness did not make a significant, long-term impact. Soon our fears of energy dependence subsided. The long lines we suffered at the gas pumps quickly became a distant memory. At the time of the 1973 embargo we were only importing 26% of our energy. Our imports today are approaching 60% and we still lack a fully implemented energy policy that could keep it from getting even worse.

The problem is exacerbated by developing countries, particularly China, suddenly becoming "gas guzzlers." The demand for energy around the world seems to be growing exponentially. And prices seem to be soaring right along with it. Over the past 30 years, the US and many countries have progressively put themselves at the mercy of energy suppliers. Countries, such as Saudi Arabia, could literally bring us to our economic knees over night—and we seem to go on as though the problem does not exist.

One bright spot in this dismal picture is the increasing realization by some that a wonderful, relatively cheap source of energy is energy efficiency. It makes what we have go further while reducing the inevitable pollution caused by burning fossil fuels.

Energy efficiency reduces operating costs and frees up funds for capital improvements. More efficient use of energy makes funds, which have been going up the smokestack, available to purchase text books, buy new medical equipment, or hire new teachers. It lowers costs for consumers, enables enhanced industrial competition, and has the potential of significantly reducing energy-related pollution. Energy efficiency (EE) can make money while reducing pollution., which makes it a very attractive remedy for some of our environmental woes.

With all the incredible benefits that energy efficiency offers us, the question inevitably emerges: why isn't energy efficiency being implemented at every opportunity? When managers around the world are asked in survey after survey for the biggest reason that they are not doing more energy efficiency work, the answer is *money*—or the lack thereof. Even in organizations, which have the requisite money to do EE work, there is always competition for the available dollars in a budget.

Unfortunately, as discussed in Chapter 1, top management seldom realizes that EE is an investment; not an expense. In fact, they seldom think about "energy" at all. A huge mental barrier exists in any attempt to sell EE to management, as management does not buy "energy," they buy what it can do. Convincing them to invest in ways to use something more efficiently when it's not even on their radar screen is a very tough sell.

Overcoming such major obstacles require considerable finesse. It helps to look at it from *their* point of view: a way to reduce operating costs that can be self-funded. Even if we get them thinking along these lines, the initial investment is still a problem. Frequently, this is compounded by a lack of technical expertise, which might otherwise make management comfortable with the concept.

There is a wonderful answer to all this: an industry that has the needed expertise, the initial capital to make it happen, and a guarantee that the work will self-fund. It offers management a clean deal of "reduced operating costs without capex." (Capex is management shorthand for capital expenditures.)

Even if the owner does not have money to invest, now more than ever, the opportunity to have a more efficient operation is great. An entire industry designed to reduce 25 percent of the owner's utility costs is here to serve their needs.

The concept is called performance contracting, a contract with payments based on performance; and the industry is referred to as the ESCO (energy service company) industry. Historically, performance contracting has been based on guaranteed future energy savings. Performance contracting allows the customer; e.g., industry, state agency, hospital, school or a commercial business, to use future savings to upgrade facilities and cut operating costs *now*.

An ESCO, which provides this guaranteed performance, will inspect a building or industrial facility for energy saving opportunities, recommend efficiency measures, and implement those measures acceptable to the owner at no up front capital cost to the owner. The ESCO then guarantees that the value of the savings will cover the cost of services and capital modifications provided the cost of the material saved, such as water or energy, does not go below a specified floor price level.

PERFORMANCE CONTRACTING IN RETROSPECT

Performance contracting is not new. The whole idea started over 100 years ago in France with a focus on district heating efficiencies. Royal Dutch Shell saw the concept's potential and exported it to the UK and the US. Today, the performance contracting concept has gained traction around the world.

Initially, Royal Dutch Shell, through its subsidiary called Thermal Energy, offered the customer equivalent conditioned space for 90 percent of its utility bills and made its profit by delivering the energy and its efficient use for less than the remaining 90 percent. From this effort at Hanneman Hospital in Philadelphia, the idea of shared savings was born in North America. Most of these early projects were based upon each party sharing a predetermined percentage of the cost savings generated by retrofits. During the life of the contract, the ESCO expected its percentage of the cost savings to cover all the costs it had incurred, plus delivering a profit. This concept worked quite well as long as the energy prices stayed the same or escalated.

But in the mid 80's, prices dropped and it took longer than expected for an ESCO to recover its costs. With markedly lower energy prices, paybacks became longer than some contracts. Firms could not meet their payments to suppliers or financial backers. ESCOs closed their doors; and in the process, defaulted on their commitments to their shared savings partners. "Shared savings" was in trouble—and the process became tainted by lawsuits and suppliers' efforts to recoup some of their expenses while facility managers valiantly tried to explain losses previously guaranteed.

To make matters worse, it was discovered that one of the pioneers in the US, Time Energy, had been entering into shared savings with an eye toward benefiting primarily from federal investment tax credits and energy tax credits. The building owner did not necessarily receive any energy cost savings. Further, Time Energy's problems with the SEC got considerable press coverage.

The industry was plagued with horror stories and soon the trust vital to accepting a new concept, was badly shaken.

Fortunately, many ESCOs persisted in their efforts to make the new concept work. Some agreements continued to show savings benefits to both parties. Of even greater importance, companies had guaranteed the savings and made good on those guarantees.

In spite of this tenuous start, the "shared savings" industry in the US survived, but its character changed dramatically. Those supplying the financial backing and/or equipment recognized the risk of basing contracts on energy prices. With uncertainty in the industry and greater uncertainty in energy pricing, risk levels grew and interest rates went up accordingly. The use of true shared savings agreements shrank to approximately 5 percent of the US market.

In its place, new names, new terms, new types of agreements, and different financing mechanisms emerged. Perhaps to respond to the negativity that had been generated around "shared savings," the industry focus turned to guaranteed performance. The term, *performance contracting,* emerged as the favored name and a new model called guaranteed savings emerged. In Europe, energy performance contracting (EPC) became the popular term for the concept, but the model remained heavily focused on shared savings.

From its shaky beginnings in the US to its near death when oil prices plunged in 1986, a strong performance contracting industry emerged in the US and has shown strong growth around the world. Two major conferences in 2005, the 1st Asian ESCO Conference in Bangkok and the 2nd European conference, called ESCO Europe 2005, are indicative of the concepts growing popularity. Some of this growth was documented in 2005 in a report by the European Commission—DG JRC, *ESCOs in Europe.**

At the turn of the century, performance contracting was blighted by "WISHCOs" entering the market. The WISHCO problem was often traced to utilities, which entered the business for all the wrong reasons. Then, utilities proceeded to dig the hole even deeper by having their unregulated subsidiaries use the same cost-plus accounting procedures that historically have given the monopolies a free ride under the federal Public Utility Regulatory Policy Act (PURPA) law. It is probably the only industry that

*Rezessy, Sylvia and Paolo Bertoldi, "ESCOs in Europe." Published by the European Commission JRC, Ispra, Italy. 2005.

can show a profit by buying new office furniture. This kind of thinking prompted a major west coast utility to build its ESCO to 415 people with only a handful of projects.

When utilities finally figured out the mess they were creating and bailed, many in the marketplace felt the whole industry was going under. Fortunately, some of the larger, more stable ESCOs, including a couple of utility ESCOs, renewed their efforts to deliver a quality product.

Performance contracting continues to change and evolve. Services are more sweeping and financial models are more flexible. As described in Section III, ESCOs today offer a broad range of retail energy services, including

- engineering feasibility studies
- equipment acquisition and installation
- load management
- supply; power marketing
- facilities management and water management
- outsourcing options
- risk management
- automated meter reading
- indoor air quality services
- energy information management
- environmental compliance
- guaranteed results

Following the patterns we have witnessed in the telecommunications industry, ESCOs are showing an increasing tendency to unbundle, and bundle, these services ... offering several or all of the above to their customers. Ultimately, ESCOs are apt to be selling conditioned floor space, which will enable ESCOs and end users a more effective and efficient means of guaranteeing the return on investment. This advent will bring us full circle back to the first US performance contract at Hanneman Hospital: chauffage—an integrated supply / use efficiency solution.

We are entering a new era where we will have a blending of energy supply and its efficient use delivered by one entity.

Further, we are apt to see the concept of energy performance contracting extended to other areas. For example, industrial suppliers are considering ways to differentiate their offering by providing performance guarantees. Several ESCOs now offer water management on a performance contracting basis. Performance contracting has already been extended to encompass operational savings (begun in the UK)

Performance contracting will work in any setting where one can define the parameters, establish a baseline and deliver the service cost-effectively for less than the baseline.

In the meantime, those just now considering performance contracting—as a consumer, a financier, or an engineering company—have a history rich in experience to draw upon. We now have the ability to look at projects, which have repeatedly put the theory of performance contracting into practice. We have learned what works. More importantly, we have also painfully learned what doesn't!

How ESCOs Work

Through the years, experienced energy service companies have come to recognize that performance contracting is primarily risk management. As a consequence, much of their work as well as the structuring of their deals focus on effective risk management and mitigation strategies.

The first critical step, then, is the qualification of customers where much of the risks lie. Attractive customers* must offer more than savings opportunities; they must have the qualities needed for a successful long term partnership. The criteria include organizational characteristics, such as business longevity and facility factors. The exact criteria will, of course, vary among ESCOs, but the process of customer selection is viewed by all successful performance contractors as a critical first step.

Since greater risks mean higher interest rates, which means the cost of money eats away more of the investment and leaves less for actual project work, it follows that it is in the customers' interests to hold risks down as much as possible. Clearly, it pays for customers to understand how ESCOs function and how to work with an ESCO to reduce risks. While this book addresses most issues from the ESCO perspective, it offers valuable insight to customers and financiers as well.

*Energy management and ESCO selection from the customer's perspective is briefly discussed in Chapter 4 and presented more fully in *Manual for Intelligent Energy Services,* which is referenced at the end of this chapter.

STRUCTURING THE DEAL

Once the customer has been selected and the general concept is agreed to, ESCOs review with the customer the services available. Once the "package" parameters are determined, the deal is structured typically in the following pattern:

1. A Planning Agreement is signed
2. An investment grade energy audit* is performed
3. Owner approves measures to be implemented
4. An energy services agreement (ESA) is signed
5. The project is implemented under the oversight of the ESCO's construction manager
6. The ESCO's project manager oversees project for the life of contract
7. O&M measures are performed as stipulated in the ESA to assure savings
8. Measurement and savings verification procedures are performed
9. Payments are made to ESCO and financier
10. Annual reconciliation is done to determine savings have covered the debt service obligation. If not, the ESCO cuts a check for the shortfall.

Financial Models

Financing models for performance contracts range from manufacturers financing a piece of equipment on a "paid from savings" basis to a sophisticated business solution. Each progressive step up the value chain, as noted in Figure 2-1, increases the complexity of the offering. At the same time, it offers greater value to the customer and more business for the ESCO.

The two dominant performance contracting models in the world, as mentioned earlier, are shared savings and guaranteed savings. The first model was developed in France and is generally credited to Compagnie Generale de Chauff. With some minor modifications, this model is the

---

*An *investment grade* audit is critical to ESCO and project success. This topic is treated in greater detail in Chapter 11. A book on the topic is referenced at the end of this chapter.

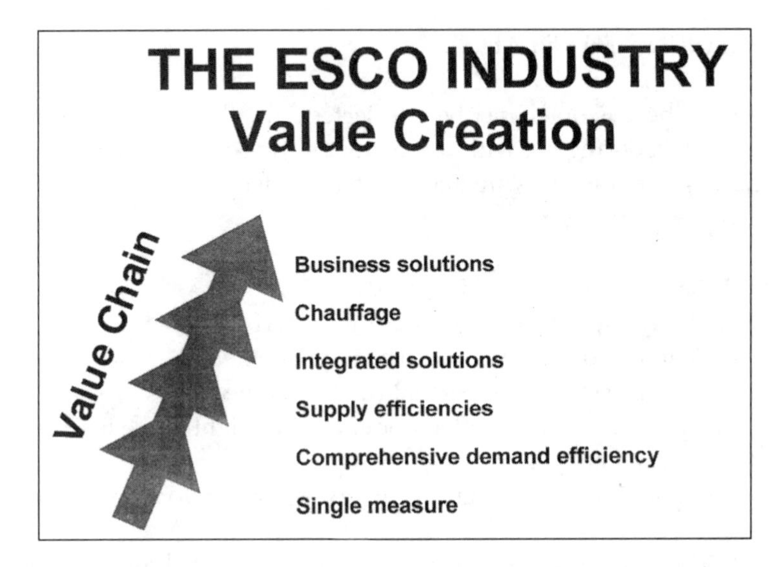

Figure 2-1. ESCO Options.

shared savings model currently practiced in Europe and, to a lesser extent, in the North America today. For about 10 years, shared savings was the only type of performance contracting offered by US and Canadian ESCOs.

The primary characteristics of shared saving are:

- Customer and the ESCO share a predetermined percentage split of the energy ***cost*** savings;
- ESCOs typically carry the financing; i.e., credit risk;
- Financing for the customer is often off balance sheet;
- Equipment, which is often leased, is "owned" by the ESCO for the duration of the contract (but ownership may be transferred to the customer at contract end)
- ESCOs carry both the performance risk and the credit risk:
- Increased risks cause the cost of money to be higher; and
- Unless special safeguards are implemented, customers have greater payment exposure if energy prices or savings increase.

The economic viability of shared savings rests on the price of energy. As long as energy prices stay the same or go up, the program will typically pay for itself.

As noted earlier, the mid-1980s saw energy prices drop, which prompted the development of a model that no longer relied so completely on the price of energy to establish the project's economic liability. ESCOs in North America shifted to guaranteeing the amount of ***energy*** that would be saved, and further guaranteed that the value of that energy would be sufficient to meet the customer's debt service obligations as long as the price of energy did not fall below a stipulated floor price.

The significant characteristics of guaranteed savings are:

- The amount of ***energy*** saved is guaranteed;
- Value of energy saved is guaranteed to meet debt service obligations down to a stipulated floor price;
- Owners carry the credit risk;
- Risks to owners and ESCOs are less than with shared savings
- Less of the investment package goes to buy money; and
- Tax-exempt institutions, in countries that provide for this tax provision, can use their legal status for much lower interest rates.

While shared savings remains the dominant model in Europe, in the US over 90 percent of the performance contracts are currently structured for guaranteed savings with the owner typically accepting the debt through third party financing.

The typical cash flow of these two financing models is shown in the following figure. In analyzing this cash flow, there are two distinguishing characteristics that should be noted.

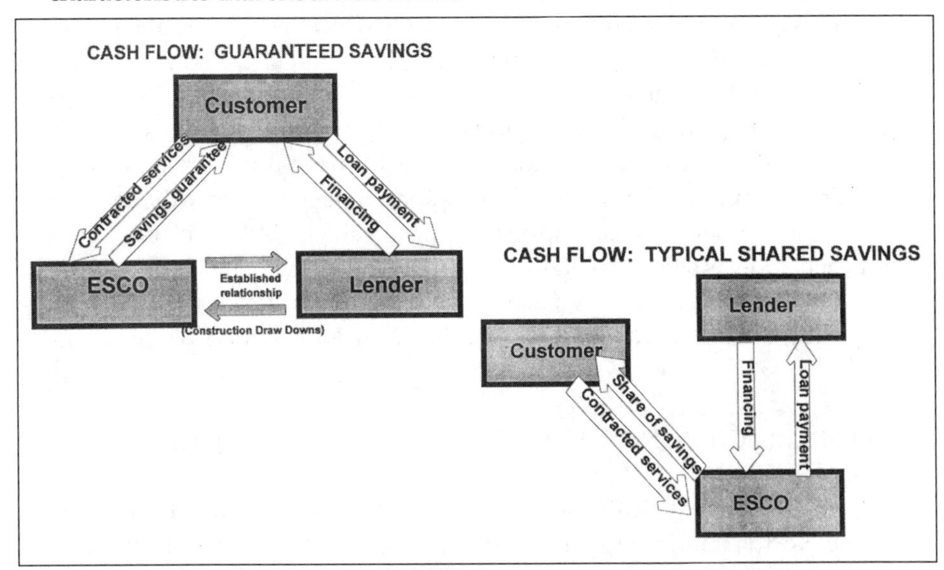

Figure 2-2. Cash flow in guaranteed and shared savings financing models.

First, in guaranteed savings the ESCO and the lender seldom have a legal relationship. Usually, an informal relationship is established and certain conditions are understood, these conditions usually involve:

- Customer pre-qualification criteria;
- Project parameters;
- Stream-lined lending procedures, which have been cooperatively developed; and
- Special interest rates.

The second distinguishing characteristic appears in shared savings. In this case, the customer has no relationship with the financing institution and has little or no interest in seeing that the loan is paid. Since all the savings must happen in the customer's facility and/or process, this factor further raises the risks to the ESCO and the financier.

Reasons do exist to temporarily encourage the shared savings model. One major reason is the difficulty customers in transitional economies have in satisfying the bank's criteria for creditworthiness. Another reason is the fact that a new concept, such as performance contracting, is easier to establish in a country if the customer does not have to incur debt. A third reasons is the desire on the part of some customers to avoid incurring further debt, or going through the political/legal procedures to do so.

Shared savings, however, relies heavily on ESCO borrowing and this presents a serious difficulty for small ESCOs which lack financial resources. After incurring debt on several projects, the small ESCO is apt to find it is too highly leveraged to obtain financing for another project. This hampers industry growth. To satisfy the small ESCOs' needs and to continue to avoid an untenable debt load, some ESCOs have turned to an emerging financial model which establishes a special purpose entity (SPE), or in some countries it is called a special purpose vehicle (SPV). The cash flow for this model is shown in Figure 2-3 on the following page.

In this model, the SPV collects the revenues and pays the financier. Typically, the financial house and the ESCO are joint owners of the SPV.

COMPARING THE NORTH AMERICAN AND FRENCH MODELS

Figure 2-4, which appears below, provides a side-by-side comparison of the North American (guaranteed savings) and the French (shared savings) models. A unique characteristic often found in the French model, as

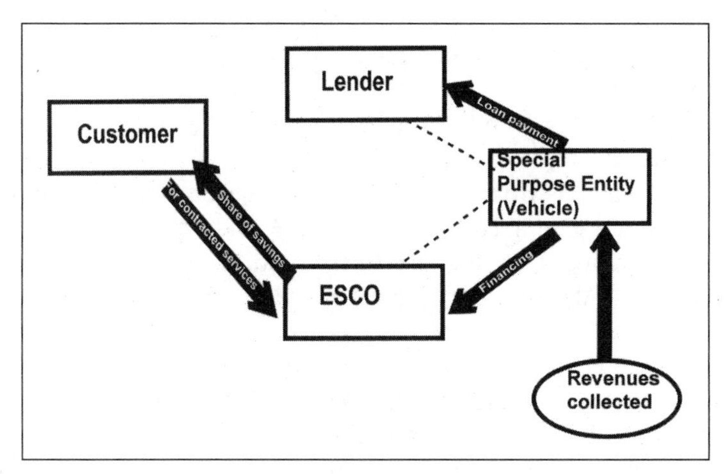

Figure 2-3. Special Purpose Entity Model

noted in the figure below, is the four-step approach. This is worth calling attention to for the French four-step procedure has two significant advantages: 1) it offers a pure self-funded approach; and 2) the opportunity to know and work with the partner before much has been invested reduces risks significantly.

This four-step approach, often referred to in North America in the 1970s as "roll-over" financing, typically uses the following pattern:

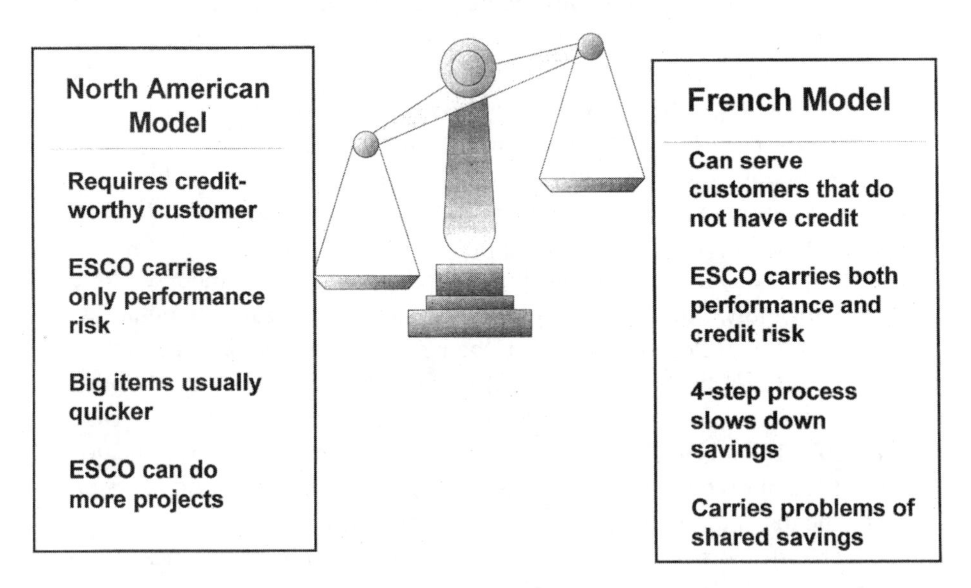

Figure 2-4. Comparing North Amercian and French Models

Step 1. energy efficient operations and maintenance (O&M) measures are implemented;

Step 2. savings from O&M measures fund the quick fix (low investment) items;

Step 3. savings from O&M and quick fix measures fund mid-range items (moderate capital costs and paybacks); and

Step 4. savings from the first 3 steps fund the "big ticket" items (costlier measures and/or longer paybacks).

As noted above, the risk management benefits provided by the four-step method are desirable. In addition, the projects generally do not require any outside financing. Under this approach, however, the end user must wait longer for major pieces of equipment and potential savings are lost in the interim. For the ESCO, the savings stream is slower to materialize and the projects are usually smaller.

MOVING UP THE VALUE CHAIN

The dominant financial models are typically applied to comprehensive demand-side management at the second level in the value chain as shown in Figure 2-1. Generally depicted at the next higher level in the value chain is supply efficiencies, which might be district heating efficiencies or the implementation of stationary fuel cells. It is placed above comprehensive energy efficiency services only because the dollar amounts can be greater for work on the supply side of the meter.

A broader range of supply acquisition services are provided, such as cogeneration or distributed generation, the package is referred to as an integrated solution.

The terms, integrated solutions and chauffage, are sometimes used interchangeably, but chauffage generally refers to a greater value-added approach. Integrated solutions may be a supply contract and a demand contract offered by the same ESCO, while chauffage offers conditioned space at a specified price per square foot (or square meter). In such a case, the ESCO manages all supply and demand efficiencies. It may include some type of ownership of the HVAC system by the ESCO, and often offers a means of adjusting energy prices on an annual basis.

The ultimate value-added on the supply chain is the business solutions approach. Typically that approach allows an ESCO to propose solutions that make prudent business sense, which may go beyond reduced energy consumption. The ESCO may provide services beyond energy efficiencies, wherein the energy cost savings may help defray the costs of this additional work. In other instances, the work may actually increase energy costs, but lower the energy cost per unit of product through process efficiencies. For example, the carpet curing procedures as discussed in Chapter 12 describes a situation where the energy costs went up; but with doubled production, the end result was a lower energy cost per unit of production.

PUSHING THE ENVELOPE

Clearly, the "bottom-line" encourages the **commercial world** to pursue energy efficiency, whether the business is a manufacturer, industrial complex, commercial offices, or residential properties. In his book *Lean and Clean Management*, Joseph J. Romm states: "Energy savings are bottom-line savings: Depending on profit margins, that $5000 [...in energy savings] could be better than a $100,000 increase in monthly sales, which entails increased costs for materials, labor, production, and overhead."

At the **local government level**, there are already more than 4,000 performance contracts in place to assist public school systems in controlling energy costs and reducing energy consumption. Since some school systems have more than 800 facilities, the performance contracting investment in reducing energy costs in public schools has offered a huge benefit to tax payers.

At the **state government level**, states continue to enact new, and improved legislation that encourages savings in public facilities and identifies performance contracting as an acceptable method of financing the projects.

On the **federal government level**, the National Energy Conservation Policy Act, as amended by further energy laws, and the Energy Policy Act of 2005 continue to require government agencies to reduce the consumption of energy. To meet these goals, the Federal Energy Management Program of the U.S. Department of Energy has actively helped agencies use energy saving performance contracting (ESPC).

Internationally, the U.S. Agency for International Development has been particularly active and effective in making transitional economies aware of the advantages an energy service company (ESCO) industry can bring to their transition to a market economy.

The multilateral development banks, such as the European Bank for Reconstruction and Development, the World Bank, the InterAmerican Development Bank and the Asian Development Bank, have all taken steps to foster and encourage the use of ESCOs in developing countries.

The expanding horizons of performance contracting are not limited to energy. **Water resource management** has found a home in the performance contracting arena. Water savings can be the basis of a performance contract, or in the case of an early example at Boston University, it can bring down the aggregate energy/water paybacks, which make the needed energy work economically viable.

A look at the numbers for the work at the university reveals the value of such a combined effort. The work progressed on a building by building basis, but when they got to Mugar Memorial Library the mechanical work, calculated at 7.62 year payback, would not meet the basic financing criteria of an aggregate payback of six years or less. H_2O Matrix joined the ESCO team and identified water retrofits; i.e., low-flow toilet and urinal flush valves, replacement of all faucet aerators with low-flow models, and the repair of domestic water leaks. The addition of these water conservation measures with a combined 2.82 year payback resulted in a cumulative project payback of 5.6 years, satisfying the base criteria economics.

The opportunities also exist for creative performance contracts that result in business solutions in **wastewater systems**. A manufacturing facility in the Northeast needed improvements to their water and wastewater systems, along with significant improvements to their production facilities to increase capacity and lower production costs. By outsourcing their water and wastewater systems, the manufacturer was able to obtain capital, reduce capital requirements, reduce operational costs, and reduce management time on the water and wastewater systems.

In this case, the outsourcing partner, Lombardo Associates/ERI Services, purchased the existing wastewater treatment facility, installed a water reuse system, and provided facility capital improvements. The manufacturer was guaranteed a reduction of $100,000 annually in operations and maintenance. The outsource partner was able to pay for the system, make improvements and lower the cost of O&M through the energy efficiency measures.

Performance contracting guarantees savings, and produces results. It is both a simple concept and a complex process. If both parties take the time to understand the options and procedures, negotiate a fair contract, and exercise the necessary commitment, it **WILL** work. There will be satisfaction.

And that's guaranteed!

Reference

Hansen, Shirley J. and James W. Brown. *Investment Grade Energy Audits: Making Smart Energy Choices,* 2004. The Fairmont Press, Lilburn, GA.

Chapter 3
Money Matters

Performance contracting is primarily a financial transaction. Return on investment is usually the motivating force for the end-user, the energy service company (ESCO), and the financing source.

Conversely, the dollars that will be lost if management doesn't act, should be a major factor in setting financial priorities. A key consideration is how much the organization will spend if it doesn't reduce energy consumption.

Understanding the financial implications of all energy actions, or lack thereof, are key. Bosses have fired energy managers and managers have become furious at ESCOs—all because the organization wasn't achieving predicted "savings," when, in fact, energy savings had been eaten up by increases in the rate schedule. The "front office" may not understand how changing rate schedules can destroy predicted savings. The concept of cost avoidance is essential to weighing and communicating performance contracting benefits, especially when energy prices are volatile.

The guidelines for weighing cost-effectiveness, calculating cost of delay and computing/graphing cost avoidance offered in this chapter lay out the money side of energy, which are important to the end-user as well as the ESCOs. First we do the calculations and then, to use an old expression, "We will run it up the flag pole and see who salutes it."

Cost-Effectiveness

Because every business, every organization, uses a significant amount of costly energy, a wide array of energy efficiency measures are available to them. However, these opportunities must be weighed to determine which measures offer the greatest financial benefit. This requires not only an evaluation of the cost-effectiveness of the measures, individually and in combination, but a broader financial analysis as well.

Ideally, major modifications will be reviewed by a committee constituted by management so implications for particular programs or impact on the work environment are part of the decision-making process. Indoor air quality and other environmental concerns should be considered. At the very least, a cost/benefit analysis requires the sharing of information and techniques between the organization's business officer and the facility manager. If the organization has an energy manager, he/she should definitely be involved in all energy-related meetings and decisions. The group should collectively weigh all the energy efficiency measures in the light of the financial benefits to the energy program and to the total operation.

Time was when cost-effectiveness ruled nearly every energy efficiency decision. Payback was the critical parameter in auditing. Today, improving the work environment and determining which measures will bring the greatest value to existing assets have become important considerations.

Two tin cans and a string are probably still more cost-effective in short distances than a cell phone, but few would opt for such an alternative. We have "tin can and string" opportunities to save energy, but they seldom add much value to the customers' physical assets and are not apt to have the useful life owners want.

Cost-effectiveness as a measure of energy efficiency is an important decision criteria and must be understood, but it is no longer the single decisive consideration.

Other operation and facility considerations also need to be considered. A new roof is not apt to be the most cost-effective measure. However, if the old roof is leaking, a new roof with increased insulation may be the most critical need. Or, replacing an old boiler that is not only inefficient, but is unpredictable and demands a lot of maintenance, may take precedence over a more "cost-effective" controls option.

Sources of funding and reimbursement implications must also be considered. Some utilities may still have rebate programs, for example, to encourage the use of certain efficient technologies.

Cost-effectiveness is one measure of economic feasibility. It is an essential ingredient in performance contracting. It answers the question: "How soon can we get our money back from this investment?" There are various ways to calculate the time necessary to recoup the cost of the original investment. These range from simple payback and adjusted payback to the more complicated life-cycle costing (LCC).

SIMPLE PAYBACK

Quick, simple and universally understood, simple payback calculations generally provide sufficient data for low to modest investments. It can also provide a good "first cut" on larger investments. Its purpose is to determine when the funds invested in a particular project will be recovered. The simple payback period (SPP) is found by dividing the value of the initial investment by the projected annual energy savings. SPP is usually given in years and/or tenths of a year.

The formula for simple payback is:

$$\text{SPP (years)} = \frac{I}{\text{ES/year}}$$

where,

SPP = Simple payback period
I = Initial investment
ES/year = Projected annual energy savings at current prices

The simple payback calculations may be modified by any factor management finds critical. The more common adjustments to be factored in are; (1) changes in operations and maintenance (O&M) costs, or (2) projected changes in energy costs. The costs of mitigating risks that might impede savings also must be considered.

SIMPLIFIED CASH FLOW

For organizations that closely follow a fiscal year budget, cost-effectiveness may be calculated in terms of Simplified Cash Flow (SCF). In this case, the computation is usually calculated within the parameters of a given period of time, often a fiscal year. SCF weighs the difference in the cost of the fuel consumed plus the difference in O&M costs against the investment for a given time period.

The formula for Simplified Cash Flow (SCF) is:

$$SCF = (E_n + O\&M_n) - (I_n)$$

where,

SCF = Simplified cash flow
E_n = Energy cost savings for the time period
$O\&M_n$ = Operations and maintenance savings for the period
I_n = Initial investment prorated
n = Period of analysis

LIFE-CYCLE COSTING

Incorporating all costs and savings associated with a purchase for the life of the equipment is increasingly being used as a means of judging cost-effectiveness. This approach, Life-Cycle Costing (LCC), may appear to administrators in government to be the antithesis of the required low bid/first cost procurement procedures. If specifications call for LCC as a means of determining cost-effectiveness, then LCC can be compatible with low bid procedures.

LCC's rather rigorous approach can be quite time consuming. For larger purchases and/or for relatively limited capital, the effort is usually justified. Life cycle costing addresses many factors which an adjusted payback analysis may miss—salvage value, equipment life, lost opportunity costs for alternate use of the money, taxes, interest, and other factors.

The simplest mode of analysis for LCC is:

$$LCC = I - S + M + R + E$$

where,

LCC = Life-Cycle Cost
I = Investment costs

S = Salvage value
M = Maintenance costs
R = Replacement costs
E = Energy costs

LCC is the net benefit of all major costs and savings for the life of the equipment discounted to present value. A building design or system that lowers the LCC without loss in performance can generally be held to be more cost-effective. Other considerations, such as the calculation of present worth, discounting factors and rates, and LCC in new design, need detailed analysis and are more fully discussed in a number of manuals and tests. The US Department of Commerce and the Federal Energy Management Program at the US Department of Energy have both prepared LCC resources in the past and remain a good reference source.

In considering cost-effectiveness, especially where guarantees are involved, the reader is referred to the discussion on investment grade audits as related to risk management in Chapter 11. While addressed in that chapter, it is important to stress here that cost-effectiveness cannot be assigned to measures without consideration of the conditions within which the measure must operate, including the capabilities of the operations and maintenance staff, the condition of existing energy-related equipment, and the costs associated with mitigating identified risks.

Cost of Delay

Some things can be put off without a loss of revenue. Energy efficiency work cannot.

Every tick of the clock, every day that passes, represents dollars an organization may have wasted by consuming needless energy. Every hour of delay forces an owner to give money to the utility that, through energy management, could have been used to educate students, train sales reps, offer patients additional services, launch a media campaign, meet constituent needs, make a bigger profit, etc.

The risk-adverse tend to sit on the funds, which could yield great benefits. Others see investments in fuzzy "future savings" as risky as the roll of the dice. Resistance from either camp just makes our job harder.

Many administrators tend to treat the utility bill as an inevitable cost. Others find that organizational pressures, which require immediate

attention, push energy concerns aside. Convincing top management and others of the immediate need to perform energy efficiency measures is discussed in Chapter 1.

But, every day you don't act is a day of wasted energy and, what is more important, a day of wasted money. For example, in Florida, a large state agency with more than fifty sites prepared to enter into a performance contract with the support of the State Energy Office. Despite previous performance contracts for state agencies, a well-meaning attorney (giving him the benefit of the doubt) took issue with the concept that a state agency would be committing itself to payments to a financial institution, regardless of the status of the ESCO or the performance of the equipment. This one misguided state official delayed the signing of a financial agreement for over eighteen months, at a cost of delay of $106,800 per month in lost energy savings! A loss of at least $1.9m to the agencies—and to Florida tax payers.

Those working in energy management have become accustomed to weighing options by calculating cost-effectiveness as discussed above. The rapidity with which energy savings recover initial investments should be a major factor in weighing energy retrofit *vis a vis* other investments. It is not unusual to find a school district, for example, with a reserve fund earning less than 7 percent interest while energy investments with an ROI of 25 percent or more go begging. Every decision maker in an organization needs to understand fully the "earning potential" in energy efficiency and the lost revenue inherent in delay.

The cost of delay is almost the mirror image of the simplified cash flow formula. The same factors that contribute to energy saving cost/benefit analysis affect cost of delay calculations, but in a negative sense.

To repeat, the SCF formula is:

$$SCF = (E_n + O{+}M_n) - (I_n)$$

where,

SCF = Simplified cash flow
E_n = Energy cost savings for the period
$O\&M_n$ = Operations and maintenance savings for time period
I_n = Initial investment prorated
n = Period of analysis

By totaling the differential costs (the anticipated savings) and subtracting the prorated investment, the positive cash flow for a specific cost-effective energy efficiency measure is determined.

The same formula can be used for Cost of Delay (CoD) where the lost savings potential becomes the differential. The lost savings differential is then reduced by the outlay that would have been needed in a given year to achieve those savings, the prorated investment. This negative cash flow figure represents the cost of postponing energy work.

$$CoD = -(E_n + O{+}M_n) + (I_n)$$

where,

CoD = Cost of Delay
E_n = Energy cost savings for period
$O\&M_n$ = Operations and maintenance savings for period
I_n = Initial investment prorated
n = A specified period of time

For example, suppose the city government is just starting an energy program and has been advised it can reduce consumption by 25 percent through energy efficient O&M procedures, low-cost retrofits, and the installation of an automated energy management control system. With an annual utility bill of $1.6 million, the avoided costs could be $400,000 per year, less the cost of work. If a $1,400,000 installation prorated over five years (at $280,000 per year) could save $400,000 per year in energy and operations and maintenance costs, the SCF would be $120,000 per year. On the other hand, no action represents a CoD of *minus* $120,000 per year.

Even when using internal resources, calculating the Cost of Delay related to various financing options makes good sense. Limited resources for other organizational needs may, in the long run, prove to be more costly. If those needs are paramount and the budget is tight, then getting the work done through performance contracting becomes an even more viable option.

An organization is frequently reluctant to use performance contracting, because it prefers to do the work itself and "save the service costs." Reality seldom meets expectations. Since O&M work is so very cost-effective, organizations often aspire to doing this work themselves so they can reap the savings before awarding a performance contracting. Unfortunately, the O&M work has a way of getting postponed and delayed, often for years, and is apt to far exceed the potential "do-it-yourself" savings.

When an organization is considering the use of in-house labor and funds to save money, it should carefully weigh the Cost of Delay first. And the delay period should be based on how long it took to get similar work in place in the past—from conception to acceptance of the installation—and not what someone hopes might be the case this time.

In addition to the cash flow impact, it pays to also recognize the benefits to the organization in improving its capital stock. Not only is the capital stock value enhanced, but the company's overall fiscal condition benefits, and typically operations and maintenance costs are reduced.

Cost Avoidance

Rising prices can wipe out all the dollar gains that have been made by reducing energy consumption. The key is to identify what an organization would have been paying if it hadn't cut back! In order to communicate energy management benefits to others as costs rise, its important to be able to talk about what would have been—the costs avoided.

The joy of counting the dollars that would have gone to the utility makes cost avoidance very real and very gratifying. In order to calculate cost avoidance, a base year must be established. The base year consumption multiplied by the current price per unit will reveal "what it would have cost."

Cost avoidance is what it would have cost minus current costs. Most top management or board members seldom have the time or inclination to wade through a pile of numbers; so it will pay to graph the data when cost avoidance for more than one year is involved.

As an illustration, Figure 3-1 depicts a cost avoidance analysis for the Great Eagle Complex. The complex had cut electrical consumption by 1,002,660 KWh since 1998, but experienced a rate increase of $0.042/kWh over the 1998 to 2006 time period. Even though consumption dropped 25 percent during this time period, costs still rose about $73,000. Without energy efficiency measures that cost would have been over $ 3.7 million. The top line in Figure 3-1 depicts what it would have cost if consumption had remained at the 1998 level, the middle line indicates actual costs of $2.8 million. The shaded area between the top and middle line shows the avoided costs of nearly one million dollars. The bottom line shows the decline in consumption.

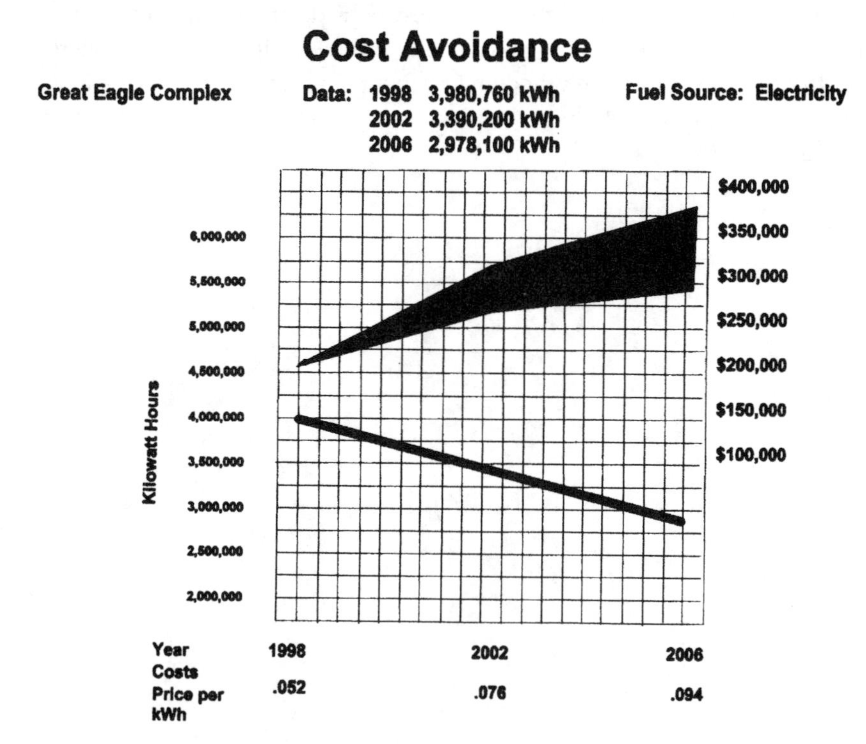

Figure 3-1. Graphing cost avoidance

The same type of graph as shown below could be used for all the fuels used in a building if total Btu is placed on the vertical axis.

Cost-effectiveness, Cost of Delay and Cost Avoidance should be critical components of energy decision-making process whether an organization plans to do its own retrofit work, or ask a performance contractor to provide the service.

ESCOs need to employ these concepts as part of their marketing technique. Cost of Delay is particularly effective. It is also imperative that ESCOs take the time to keep customers informed of their cost avoidance and cumulative benefits over the life of the project. It is recommended that such information accompany any invoices for services to the customer.

Cost avoidance numbers, preferably graphs, should be a regular feature in the monthly billing process. Without it, customers quickly forget what the utility bill used to be and why they are paying the ESCO. This becomes particularly critical when there has been a change in administration or top management.

Finally, it is important to never lose sight of the total facilities/processes as part of the organization's investment portfolio. New equipment, modifications, and energy services should all be designed to improve that portfolio and enhance the work environment. Far too often, we have let the tail wag the dog and let the cost-effectiveness of measure be the sole determinant.

Chapter 4

Partner Selection: From Both Sides of the Fence

When entering a performance contract, the biggest risk, for ESCOs and owners alike, is usually the selection of the performance contracting partner. It is not an exaggeration, therefore, to say that for many this is the most important chapter in this book.

A common lament from energy service companies, and those of us who help customers secure ESCO services, is, "Please figure out what you want to do first *before* the request for proposal (RFP)* is issued." Second-guessing and changing intent in mid-stream create headaches for all—and bigger price tags. The costs to ESCOs, the frustrating time delays, the owner's cost of delay, the procedural (and legal) problems are legion.

These BIG headaches of eye-blurring migraine proportions are totally unnecessary and avoidable. A little thought and planning can make all the difference.

PREPARATION: FROM THE OWNER'S SIDE

First, a key admonition: do your homework. For the owner, a few "Homework Rules" will help avoid costly delays and hugely expensive disasters.

Rule #1 for any owner seeking ESCO services is to: DETERMINE THE RESULTS YOU WANT FIRST. Gather a good cross-section of people in your organization together and decide what *RESULTS* you want out of a performance contract. Since performance contracting rests on technical, legal

---

*We have bowed to the inevitable in this discussion and refer to all requests for ESCO services as RFPs, which is common in the market place. As noted later in this chapter, there is rarely a need for a full blown RFP and a request for qualifications (RFQ) will usually serve.

and financial procedures, the team should include technical, legal and financial people. And be aware their tentacles can reach far—into the project and into your organization.

A quick aside: one major value of the first meeting is the opportunity for everybody to air a lot of apprehensions and even fears. The flip side of these apprehensions is positive results. If, for example, somebody expresses concern about negative impact on the indoor air quality, the results you want might be framed in air changes per hour or in CO_2 parts per million.

Many of the problems related to securing ESCO services are rooted in the idea that performance contracting is a technical solution; so someone in management thinks the RFP process should be handled by the facility people. Of course, facility management must be involved. Sound technical services is a key component. But performance contracting is primarily a financial transaction and even the very best facility people should not be asked to go it alone.

The owner (and ESCO for that matter) should, at the outset, recognize who in his or her shop makes the decisions and who influences those decisions. Energy managers and facility managers can be very influential and may easily kill a deal, but they seldom sign the contract. Those who will ultimately okay a contract (lawyer) and sign a contract (CEO &/or CFO) should be involved from the beginning. If they are not represented in the process from the beginning, chances increase dramatically, even exponentially, that major procedural changes and delays will occur mid-process.

RESULTS is the by-word for Rule #1. Yes, an owner may desperately need a new boiler or an up-date in the controls system, but the goal is a more efficient, less costly operation while maintaining, or even enhancing, the work environment. The organization is buying the ESCO's expertise in reducing operating costs as well as the associated services to make it happen.

Equipment is merely the vehicle used to deliver the expertise, services, and savings. An owner, who secures an audit and then proceeds to list the needed equipment in the RFP, is not entering into an effective performance contract. Such a procedure raises the ESCO's and the customer's risks, *and costs,* tremendously. It also seriously limits the creative knowledge ESCOs bring to the project. Instead of a broad range of proposals full of energy efficiency ideas, the owner has the perspective of one engineer, who in all likelihood does not guarantee his work.

The RFP team should never lose sight of the organization's need for a cost-effective, productive work environment. The goal is not to cut energy consumption *per se* but to use the energy that must be used as efficiently as possible. Safeguards in operating parameters are essential components of the contract, and need to be reflected in the RFP. Setting acceptable heating and cooling temperature ranges, lighting levels, humidity range and air changes per hour offer occupants the assurances they want and give an ESCO the information it needs to decide whether it should even consider the job.

Often, safeguards are situation specific. A very successful ESCO, whose track record in heavy industry is outstanding, may be a disaster in a hospital. For a hospital, a major concern is infection control and *nosocomial* is a dreadful thought. An ESCO, which does not understand the use of positive and negative pressure or isorooms and has not had successful experience in a hospital, should NEVER be selected to do hospital work.

The RFP, therefore, should state the general and more specific criteria an ESCO must meet to be selected. It should lay out the fundamental results the owner is looking for and the criteria the owner intends to use to determine which ESCO appears to be most qualified to deliver those results.

Rule #2, then, for any customer is to DECIDE THE CRITERIA the organization will use to determine which ESCO can best deliver the results it wants. Setting those criteria and how they will be used in the evaluation process is discussed later in the chapter.

For *Rule #3*, we get really dogmatic. The EVALUATION PROCEDURES, which incorporate those criteria, SHOULD BE WELL THOUGHT OUT *BEFORE* THE CUSTOMER ISSUES AN RFP. Each time we have stepped in to help an owner after the RFP is issued, we swear (a lot) that we will never again go into a selection process when the RFP has already been issued. Just one really frustrating experience will explain why.

When an administrator had narrowed the potential ESCOs to four, he decided he needed our help. We foolishly agreed to help him with the process. The owner had issued a RFP titled, "Request for Proposal for Performance Contracting for Energy Conservation Services." The first criteria for selection as stated in the RFP was, "Preference shall be given to respondents who demonstrate strong capabilities in terms of experience and reputation in performance contracting." When it came time to evaluate the proposals, the owner changed the rules of the game and decided that the proposers didn't need performance contracting experience! (For reasons not fully explained, but … an ESCO with no experience was favored.) For this and other reasons, we recommended the owner close the solicitation and start over. To do anything else was a gold plated invitation for litigation. The owner's solution? The four ESCO finalists were gathered together and asked to sign a paper saying they would not sue! When we expressed concern about this plan, we were told, "We do things differently in Texas." We quickly walked, no make that *ran* away, from any association with their whole process.

Once the preliminary homework has been completed, it's time to consider writing the RFP. Even though Oliver Wendell Holmes is said to have observed, "No generalization is worth a damn, including this one," we will generalize that RFPs are inevitably longer than they need to be and ask for far more information than most owners need. Or, know what to do with. And so we come to Rule #4.

Rule #4 for soliciting proposals is pretty simple:

- KEEP IT SHORT!
- KEEP IT OPEN!
- KEEP IT SIMPLE! AND
- GET ONLY THE INFORMATION YOU ABSOLUTELY NEED TO MAKE A QUALITY EVALUATION!

Full compliance with Rule #4 would cause a collective sigh in the ESCO industry and create a major drop in performance contracting headaches.

Proposals that cover all facets of energy financing and services are cumbersome and costly to evaluate and expensive for an ESCO to prepare. The time and cost of preparing a major proposal is apt to discourage some ESCOs from even submitting one, especially to smaller organizations. This is especially true if the savings opportunities are limited, or the administration or delivery of services relatively burdensome. A cumbersome solicitation process, therefore, definitely limits the options available to smaller organizations—and does not do any favors for larger organizations.

Developing An RFQ/RFP

Prior to structuring the evaluation process, the committee should meet and agree on required results, evaluation procedures and RFP wording. Preferably, the criteria, weighting, etc. will be set by the committee before the solicitation is issued. The solicitation, such as a Request for Qualifications (RFQ) or RFP, will inform potential proposers of the services required and the criteria that will be used to judge their qualifications. Assessing the organization's needs and setting criteria are discussed later in the chapter.

Some very pragmatic thoughts for RFP writers are offered below in the hope that they will make life simpler for everyone involved.

Keep it short! Extensive requests for information pro-vides an organization with a huge evaluation burden. Further, it takes time to prepare big requests for proposals (RFPs). And the owner does not do him or herself any favors by doing so. Second thoughts and a shredder may become dear friends.

Long, involved RFPs can be excused only if: 1) internal politics and procedural concerns demand it; or 2) federal agencies or some state governments must meet acquisition regulations. ESCOs with federal and state procurement specialists on staff can usually play successfully in this quagmire, but all others should proceed with extreme caution. Most RFPs (outside of government) should not exceed 30 pages, plus any appendices. Well, government ones should be shorter, too, but getting bureaucrats to change procedures is usually more trouble than its worth.

KEEP IT OPEN! Detailed specifications are a throw back to "bid/spec." Performance contracting solicitations should be based on eliciting information as to the ESCO's qualifications—not price. (There are some "price" exceptions, e.g., the federal government insists on some bid prices and the Canadians use an "open book" approach.) But bidding is delusional and typically works against the owner's needs. Good bidders know how to play the numbers game.

Expert bidders will meet specs and not offer the owner a dime more. They are experts at delivering ***barely acceptable*** products—after all, that's what bid/spec asks for!! If, as an owner, you are hoping for some post-contract savings, you are not apt to get it with bid/spec. If you are required to use bid/spec, then consider writing USEFUL LIFE requirements into your specifications.

Describe what results you want; not how to do it. If you know exactly how to do it, you probably don't need an ESCO. Detailed specifications may eliminate options that ESCOs could have offered. The spec approach means your organization is relying on one auditor's thoughts and losing out on all the rich experiences ESCOs could have made available to you. It's their BUSINESS to identify all cost-effective energy conserving opportunities; not yours.

Precise specifications not only put a high wall around the ESCO's opportunity to help; it may spell out a project in such a way that the ESCO cannot guarantee the savings. It is not unusual for owners to find "spec" RFPs are, as the head of the US Postal Service once put it, "non-responsive." Freely translated, that means no proposals were received at all.

KEEP IT SIMPLE! Some RFPs raise the question, "Who are you trying to impress?" Or, "You want what!?!" The owner is more apt to get clean, direct proposals if clean, forthright RFPs are used.

The obligations imposed on those responding to the RFP should be kept within the realm of reason. In the early 1990s, an organization with 200+ buildings issued a PRELIMINARY request for qualifications that required the responding ESCOs to audit facilities and to guarantee a level of savings for ALL its buildings. At this preliminary stage, the idea was absurd and served no purpose! Since the majority of the buildings were schools, which are seldom all that unique, the request bordered on the ridiculous. No, it surpassed the ridiculous! Then, they proceeded to demonstrate to the world that they did not have the foggiest idea what they were doing by indicating they expected a three-year contract. Since it was a large district with huge potential in over 200 facilities, some ESCOs did offer qualified responses in the hope that the school system would get more realistic as the process developed. They did get a proposal from an ESCO which saw the opportunity to sell some equipment. The ESCO talked them into a six-year contract. The results were pretty ugly and the owner fell far short of what it could have had out of the project.

GETTING THE BEST

The more you ask for, the more tedious and costly the evaluation process. Most "kitchen sink" RFPs are a testimonial to organizations which have not truly thought through their needs and/or the performance contracting process. Or, managers who must satisfy so many internal masters that the primary purpose of the document is all but forgotten. In "many masters" situations, the use of oral interviews for the pre-qualified few should be considered.

The "get only what you need" admonition cuts both ways. ESCOs should be constrained by format and length. Unless it's one of those government things, one must worry about an ESCO that can't tell you who they are, what they can do, and how they package their projects in 30 pages or less—excluding appendices and sample audits.

TO TEST OR NOT TO TEST

In the 1970s when the performance contracting process was foreign to owners, test audits were *de rigueur.* How else could the technical capability of the ESCO be judged?

Fortunately, the desire, and certainly the need, for a test audit has faded with time. Unique organizations with unusual operations or peculiar problems may warrant a test audit on a portion of a facility. All others should resist this cumbersome embellishment if they possibly can.

The major difference between a full blown RFP and what has become known as a request for qualifications (RFQ) is generally the technical requirements; and more specifically a test audit.

Owners should keep in mind that ESCOs, who repeatedly succumb to test audit conditions, have to recover their costs somewhere. The test audit costs run up the overhead. Ultimately, those costs must be borne by the customers—particularly those organizations that asked for test audits. In other words, if you ask for a test audit, you should expect to pay for it one way or another.

Today, established ESCOs have proven track records and an increasing number of organizations are accepting sample audits of similar facilities, or well documented references/case studies in lieu of on-site test audits. A test audit is just one more obstacle that may prevent a well-qualified ESCO, which could meet an organization's needs, from even proposing. Audits are costly and ESCOs are increasingly reluctant to incur the cost of an audit on speculation.

At the very least, audits should only be requested from a short list of pre-qualified ESCOs. This saves the ESCOs time, improves their odds and reduces the burden on owner's organization, which have to evaluate all those test audits.

The most attractive alternative is to ask proposers to submit an audit for a similar facility, which is representative of the caliber of work they propose to perform. This sample audit can still be evaluated by an engineer representing the owner's interests and can also be used as a basis for a representative presentation of an ESCO's financial approach. The owner should reserve the right to attach the sample audit by reference to its contract as representative of the ESCO's standards of practice. This reduces the chances of getting a "solid gold Cadillac" as a sample and a Chevy as the delivered product.

The discussion here has referenced the audit as part of the selection process. A preferred procedure, growing rapidly in popularity, is a preliminary contract called a Planning Agreement, which is entered into with the selected ESCO. The Planning Agreement specifies the criteria for an acceptable audit to the owner and requires the owner to pay for that audit if the project does not got forward. (See Chapter 7, "Quality Contracts," for more information on Planning Agreements.)

PRE-PROPOSAL CONFERENCE

The RFQ/RFP process frequently also involves a conference for prospective proposers, particularly for larger or complex projects. Much like a bidders' conference, the pre-proposal meeting is designed to clarify and expand upon the organization's needs and intent. If it has not been made part of the issued RFQ, the owner's lead person should be prepared to describe the facility(ies), operating conditions and offer utility records for at least two years, preferably three, on all candidate buildings.

As part of the conference, proposers should be given an opportunity to walk through a representative building while staff respond to questions. Even though the more exacting technical considerations may have been deferred, prospective proposers should have some opportunity to judge the organization's savings potential and whether the job potential meets *their* criteria.

Required attendance at pre-proposal conferences puts a burden on the industry and can reduce ESCO options for the end-user; so required conferences should be well structured and substantive. Unless the project is very large or complex, required attendance at the pre-proposal conference should be avoided if possible. This is just one more requirement that raises proposer's costs and discourages ESCO participation.

If the project is awarded to an ESCO that did not attend a pre-proposal conference, then a real question as to the value of the pre-proposal conference must be raised. A required pre-proposal conference, therefore, limits owner's options downstream and should be called only if there are some unique concerns or procedures that may substantially affect the proposals or the selection process.

Unless the pre-proposal conference is a required part of the selection procedure, a firm's attendance should not, technically, be a factor in making the selection. The conference does provide a subtle opportunity, however, to size up the proposers, and for ESCOs to make a favorable impression. It is also an opportunity for an ESCO to size up a candidate for its services! Since performance contracting has been compared to a marriage, the pre-proposal conference might be viewed as a the first date in a "courting ritual." Finally, a pre-proposal conference also gives the owner an opportunity to test the level of interest and the ESCOs are offered an opportunity to judge the competition.

INTERVIEWS

After the preliminary selection of qualified firms, an organization can elect to hold oral interviews. If known in advance, the intention to hold interviews should be stated in the RFP.

If oral interviews are held, the letter of invitation should outline the nature of the presentation expected and any other requirements to be placed on the proposer. To the extent possible, the ESCO should also be provided information as to who will be representing the organization at the meeting. ESCOs will often ask who the other finalists are and in what order they will be interviewed. Whether this information is supplied is, of course, totally at the discretion of the owner. A good gauge is for the owner to consider what *the organization* will gain by providing an ESCO such information.

Owners should always insist on having the ESCO personnel present who have been designated to serve the owner's project needs. Face-to-face assessments can enhance the evaluation process. It also makes it clear who will actually be on the job and which, if any, impressive names/positions were put on the proposal as window dressing.

Further, it behooves the owner to ascertain how much time certain members of the designated ESCO team will actually be on site. Having impressive people as part of the ESCO presentation team, who will do little more than walk through the facility later, occurs entirely too often. It is at the root of the term "beltway bandits," used to designate firms around Washington, D.C. who use such tactics to get federal jobs.

After the interview, many ESCIOs have found it useful to request a summary statement confirming or clarifying certain points raised during the meeting. This is particularly pertinent if new conditions or offerings have been presented in the interview process. The owner may also wish to have a "paper trail," which might avoid future misunderstandings and litigation.

A TWO-STEP SOLICITATION

Since the firm's qualifications to deliver the results are the most important single criterion, a frequently accepted practice is a brief request for qualifications (RFQ) to create a short list for the assessment of technical competence. If the owner's organization is inclined to use this approach, it should exercise care that this preliminary screening does not yield qualified firms that look good on paper, but in practice offer inexperienced teams to do the work. In most cases, however, an RFQ with a sample audit will serve the owner's needs.

While a two-stage solicitation is encouraged if a test audit is part of the process, owners should remain keenly aware that there are disadvantages to relying on a RFQ/RFP two phase effort. They are: (1) the qualifications proposal usually does not describe the ESCO's full technical approach to the project, or the financial benefits to the organization; so the evaluators will not know for sure whether they have selected the most favorable candidate (the most attractive opportunity, therefore, may be lost); (2) ESCOs may view the two phase submission as too burdensome, particularly for a smaller organization, and may be discouraged from proposing; (3) it protracts the selection process, thus lengthening the time the organization must pay for wasted energy; and (4) adds to the overall administrative burden. A two-step selection process also lengthens the sales cycle for the ESCO, which ultimately adds to product/service costs.

REQUESTS FOR QUALIFICATIONS FORMAT

Owners should ask for, and expect, a listing of who will be assigned to the job, their assigned responsibilities AND THEIR QUALIFICATIONS TO FULFILL *THOSE SPECIFIC* RESPONSIBILITIES. An estimate of the amount of time they will actually spend on the job is also key. Otherwise, you are apt to have a high powered engineer breeze through leaving Engineers in Training (EITs) in his wake to do the work.

It pays to remember the first law of RFQ writing: THE MORE YOU ASK FOR THE LESS YOU GET. Defining the results rather than the means will reduce the burden of RFQ preparation and encourage more innovative responses.

Figure 4-1 presents the basic elements of an RFQ. It's critical to remember that the basic elements only offer a framework for developing the RFQ that will meet *your organization's* particular needs. There is no such thing as a model RFQ that will suit everyone. Much is lost in the process unless the model is adapted to address the unique conditions of a given organization.

REQUEST FOR PROPOSALS FORMAT

Should an organization wish to use a broader request for proposal (RFP), the RFQ can be converted to an RFP by adding components that solicit information on the technical approach and associated financial calculations. Language eliciting technical/financial competence should set the auditing and financial parameters. Should an owner decide a test audit must be done on a given facility(ies), such language should be incorporated in the solicitation document.

Other Modifications

In converting the RFQ to an RFP, the document should be carefully

REQUEST FOR QUALIFICATIONS

1. Purpose and scope—As briefly as possible, offer the over-all purpose of the project, the range of services you are looking for, and any limitations of importance to an ESCO.

2. Proposal procedures—submission information and expectations regarding the RFQ or later submission requirements; e.g., whether an audit will be expected from the pre-qualified firms.

3. Pre-proposal—(bidders) conference information (if one is scheduled).

4. Selection criteria and any weighting.

5. Request for brief case study references which include:
   a) demographics;
   b) projected and actual costs;
   c) measures implemented;
   d) predicted and actual savings; and
   e) contact information.

6. Contract requirements pertinent to costs to be incurred by contractor; e.g., insurance requirements.

7. Proposal format, content, and preparation instructions
   a) Contractor background and qualifications.
   b) Trade references.
   c) Personnel designated to participate in the project, responsibilities, qualifications, experience in similar facilities and percentage of time each person will devote to the project.
   d) Sub-trades performed and reliance on joint venture or subcontractors; qualification information on joint venture and/or major subcontractors.
   e) Prior relevant experience and references. Ask for one reference where the ESCO had some problems; so you can check to see how they resolved the problem(s).
   f) Annual report or audited financial statements for the ESCO's most recent fiscal year.
   g) Demonstrated capability to finance the project.
   h) Demonstrated level of performance bonding; professional liability insurance.

8. Assurances, qualifications, limitations.

9. Evaluation procedures to be used; selection criteria (and any weighting to be used); and

Figure 4-1. RFQ Content Outline

reviewed for any cosmetic changes needed to make the solicitation consistent throughout. Other procedural changes, such as the time specified in the deadlines, should consider the longer time period required for an ESCO to fulfill additional requirements and/or conduct a test audit.

Whatever process is used, the key to effective solicitation procedures is to ask for the information that will enable the owner to judge the proposers' qualifications and competence to meet the organization's particular needs. A good request for proposals identifies those needs; and, if an ESCO looks closely, they are further implied in some of the details in the suggested criteria.

ESTABLISHING CRITERIA

In actual practice, the process of evaluating proposals will follow the solicitation procedures discussed earlier in this chapter. The intended evaluation procedure, however, precedes a discussion of solicitations because evaluation criteria and process concerns should influence the way the request for qualifications and consequent responses are developed. As Abraham Lincoln once observed, "If we could first know where we are and whither we are tending, we could better judge what to do and how to do it."

Evaluating the qualifications of an energy service company usually requires a multiple disciplinary approach, including technical and financial expertise. The evaluation process, therefore, typically involves a committee. Unless the organization soliciting an ESCO's services has in-house performance contracting experience, it pays to support the committee's deliberations with a consultant.

Before putting pen to paper, an honest appraisal of in-house capabilities to evaluate performance contractors' qualifications is warranted. Customers often seek performance contracts because they do not have the funds to do the needed work on their own. This frugal posture may cause resistance to hiring a consultant. Unless someone in the organization is thoroughly familiar with performance contracting, this is an exceedingly expensive way to "save money." The costs of a performance contract consultant, and/or an engineer as a technical consultant, can be assigned to the project, and the costs covered by future savings. But in fairness to the ESCO, these contemplated procedures should be mentioned in the RFP.

SCORING

To establish consistency among evaluators, scoring procedures need to be determined at the outset.

Proposals are generally scored on a 0 to 10 scale for each criteria and are entered on an evaluation form, usually a worksheet for each of the major criteria. Those scores are then transferred to a summary sheet. At this point, if the criteria are to be weighted, the total score for each major criterion is multiplied by the agreed upon weighting so it reflects the relative importance of the criterion. The sum of the weighted scores for each criterion provides the total score.

The scores are usually based on a frame of reference similar to the following:

0) Criterion was not addressed in the proposal or the material presented was totally without merit.

1) Bare minimum.

2) Criterion was addressed minimally, but indicated little capability or awareness of the area.

3) Intermediate Score between 2 and 4.

4) Criterion was addressed minimally, but some capability was indicated.

5) Intermediate Score between 4 and 6.

6) Criterion was addressed adequately. Overall, a basic capability.

7) Intermediate Score between 6 and 8.

8) Criterion was addressed well. The response indicates some superior features.

9) Intermediate Score between 8 and 10.

10) Criterion was addressed in superior fashion, indicating excellent, or outstanding, capabilities.

Worksheets for each criterion can be broken out into factors to be considered. A detailed listing of these factors makes sure that every evaluator considers each aspect—and that the proposer has not omitted, whether purposely or inadvertently, any important information.

EXAMPLE

Example of the above criteria applied to team qualifications are shown below.

CRITERION: QUALIFICATIONS OF THE PROPOSING TEAM

Factors To Be Considered:

a) Experience of the prime contractor with previous projects of similar size and type.

b) Experience of the joint venture partner (or suggested subcontractors in the team) with previous projects of similar size and type relative to their stated special expertise.

c) Experience of the proposed project manager as it relates to this project, as well as the qualifications of the assistant project manager; site manager; financial specialist; engineers for design; etc. Percentage of time suggested personnel will devote to the proposed project.

d) Resources (other than financial) available to the team for computer-aided design, equipment fabrication, test/checkout, on-site assembly/erection, commissioning training, etc.

Score:

0 = Personnel appear inexperienced; resources appear insufficient for the job; allotted time inadequate.

2 = Some personnel proposed appear marginally capable; resources appear limited; allotted time questionable.

4 = Personnel proposed appear competent in their fields; some resources are available; allotted time marginal.

6 = A majority of the personnel proposed are experienced in the general fields required; adequate resources; allotted time barely adequate.

8 = Generally experienced personnel; allotted time adequate, some unique qualities, adequate resources.

10 = Fully-qualified and experienced personnel; comprehensive facilities experience, particularly qualified to perform designated duties, sufficient time allotted, excellent resources.

Examples using the criteria for financial arrangements, technical performance and management along with sub-criteria and scoring procedures are presented in Appendix B.

THE EVALUATION PROCESS

Prior to structuring the evaluation process, the committee should meet and agree on definitions, scoring and procedures. Preferably, the criteria, weighting, etc. will be set by the committee before the solicitation is issued. The solicitation, such as a Request for Qualifications, will inform potential proposers of the services required and the criteria that will be used to judge their qualifications.

The criteria should grow out of the identified needs by a cross-section of the organization—based on a fundamental and objective assessment of in-house capabilities.

Since the details in a proposal seldom fit preconceived molds, the decision-matrices used in the evaluation format provide *only suggested criteria*. The owner needs to enter the criteria, which was developed by the committee, in the decision matrix format. The committee should then address these criteria as they independently evaluate the proposals.

The approach shown in Figures 4-2, and 4-3, offers the flexibility to highlight particularly attractive features or some strongly held reservations. The decision-matrix summary sheet and the supporting worksheets allow the evaluator to view and compare at a glance, the ways each firm treated certain criterion. For more complex projects, the decision-matrix approach is preferred.

To repeat: the criterion and weightings suggested in these figures are just that; suggestions. They tend to reflect actual practice, but every organization needs to decide the relative importance of certain criterion. For example, the weight of "5" for the proposal presentation is established to encourage proposers to follow the format prescribed in the RFP, which will facilitate the evaluation process. The object of the effort, however, is to select a firm, not a proposal. If too much weight is placed on the proposal presentation, the process could defeat the purpose.

CHECKING REFERENCES

As part of the evaluation process the customer should always ask for, AND CHECK, a potential ESCO's references regarding work they have done

Figure 4-2. Proposal Evaluation Financial Benefit Worksheet

Financial Benefit Worksheet

CRITERIA/FIRM					
PROJECTED LEVEL OF TOTAL ENERGY SAVINGS AND CAPITAL INVESTMENT					
ORGANIZATION'S SHARE (% OF SAVINGS)					
INNOVATIVE ENERGY FINANCING - Payment Schedules - Interim Construction Financing					
CONTRACT YEARS & RELATION TO - Savings - Services - Other Benefits					
FORMULA - Establishing Baseyear - Billing Calculations - Demand charges - Floor price					
BASELINE ADJUSTMENT - Occupancy - Weather - Energy prices - Operating hours					
LEVEL OF INVESTMENT IN CAPITAL EQUIPMENT & MODIFICATIONS (%)					
EXPLICITNESS AND FAIRNESS OF METHODOLOGIES					

Figure 4-2. Proposal Evaluation Financial Benefit Worksheet (*Continued*)

ESCO FEE - in guarantee pkg. - partly in excess savings - all in excess savings					
RISK EXPOSURE OWNER REQUIREMENTS - Insurance - Operational control - Guarantees - Payments for maintenance					
PROJECT TERMINATION - Buyout provisions - Return to original status					
OPERATIONAL SAVINGS CALCULATIONS - clearly documented - real budget savings - requires reduce man-power to achieve savings					
COMMENTS EVALUATOR ________	TOTAL	TOTAL	TOTAL	TOTAL	TOTAL

Figure 4-3. Decision-Matrix—Summary Sheet

Criteria/Firm				
Proposal Presentation	_x 5 = ____	_x 5 = ____	_x 5 = ____	_x 5 = ____
Firm's Qualifications	_x 30 = ____	_x 30 = ____	_x 30 = ____	_x 30 = ____
Technical/Service	_x 20 = ____	_x 20 = ____	_x 20 = ____	_x 20 = ____
Management	_x 15 = ____	_x 15 = ____	_x 15 = ____	_x 15 = ____
Financial Benefit	_x 30 = ____	_x 30 = ____	_x 30 = ____	_x 30 = ____
Comments:				
Evaluator	Total	Total	Total	Total

in similar facilities. The references cited by an ESCO, of course, are usually the best they have to offer. Consequently, it really pays to ask the ESCO for a reference regarding a project that had problems. When you talk to this customer, you will quickly find out how well the ESCO functions in adverse situations.

Several sources are available to the enterprising procurement office. Colleagues are always good sources. The state energy office in the ESCO's home state may be a viable source. The state personnel cannot endorse a private firm and are not apt to rule out any firm; however, listening to what they "don't say" can help. If they suggest you contact certain people, who have used the proposer's services, do it. Indirectly, you'll find out in a hurry whether the energy office is high on a certain firm. Or otherwise.

If you still remain uncertain as to the potential performance of a proposer, members of the committee can resort to the ultimate check-up: ask the competition in the area to identify any project where other proposing firms did not perform as promised.

CONTRACTOR SELECTION

While not essential, it is usually valuable to have the committee meet again after independent judgments have been made and submitted. This removes the possibility that committee members misunderstood the instructions, allowed biases to intrude, or overlooked a key area that might be brought out by another committee member during the discussion. Then, to make sure everything is neat and tidy, each committee member should make a clean set of worksheets that can go in the file.

After the contractor has been selected by the committee, it is generally necessary to have top management or a board ratify this action.

This ratification is generally followed by a letter of intent to the selected firm. This letter notifies the contractor of his/her firm's selection and stipulates the time frame and conditions for contract negotiations.

Typically, a Planning Agreement for the investment grade audit and/or a master contract is then established. In the typical Energy Service Agreement, specific recommendations, associated maintenance costs, etc. for specific buildings or complexes are treated in addenda or schedules. These procedures are more fully discussed in the chapter on contracts.

The evaluation and selection process for securing the services of a performance contractor is unique. Potential customers are sometimes reluctant to enter this process without outside support. Ironically, customers often seek performance contracting because they do not have the funds to do the needed work on their own. As noted earlier, the costs of a performance contract consultant, and/or an engineer as a technical consultant, can be assigned to the project, and the costs covered by future savings. It is false economy to not act due to a level of uncertainty that could be resolved by a consultant, or because of the cost of that consultant. The cost of delay easily surpasses the cost of the consultant. Since the funds to pay for the consultant can be covered from the project revenue stream, avoiding this expense does not make sense.

The most costly evaluation, selection process is an attempt to proceed with in-house staff that do not understand the needed procedures and safeguards. It is reminiscent of the old oil filter commercial, "You can pay me now; or you can pay me later." In performance contracting, "later" is a lot more expensive.

PREPARATION: THE ESCO'S SIDE

In an industry that has shown maturity in so many ways, it is absolutely amazing how many really ghastly proposals are still submitted to owners. It staggers the mind that some ESCOs think such non-responsive gibberish will actually get them jobs. (Of course, it really blows the mind because sometimes those proposals do get jobs When that happens, one must wonder if the proposal had anything at all to do with the selection.)

Writing quality proposals is not a mysterious or marginal event. Knowing how to listen to the potential customer and how to read/analyze an RFP will go a long way towards writing effective get-the-job proposals.

The above discussion of the owner's ESCO selection procedures identifies many factors that the ESCOs need to be cognizant of as they prepare proposals, including the owner's needs, which should be reflected in the proposal. Listening to the owners concerns and trying to view things from the owner's point of view is a critical first step.

When an RFP comes into an ESCO office, a quick scan will usually determine if further deliberation is warranted. Then, it's the circular file or time to get to work.

As discussed in Chapter 11, "ESCO Risks and Management Strategies," the potential customer represents a huge risk and careful procedures to evaluate a customer as a potential performance contracting partner are critical to a project's success. As this evaluation is moving forward, two other activities will be proceeding simultaneously: 1) the astute ESCO will be gathering proposal ideas; and 2) every contact with the potential customer will be viewed as an opportunity to sell the ESCO's interest in the customer's needs and its ability to serve them. These two activities should provide some tips on ways to frame the proposal/presentation/sell to differentiate a given ESCO's offering from the competition. This is an area where ESCOs frequently fall short. As one university's director of physical plant put it, ESCOs seem to be "creatively challenged."

To prepare proposals efficiently, a certain amount of "boiler plate" needs to be in the computer banks for easy access. But the time this efficiency frees up should be used to distinguish an offering from the competition, particularly with ways you can be more responsive to a given customer's needs.

PROPOSAL STEPS

A well-written RFP will clearly state the format to be used for proposal submission.

To organize the proposal preparation, it helps to follow a few simple steps.

Step 1. Build a spread sheet and put the required format headings as stipulated in the RFP in a right hand column.

Step 2. Now, read the RFP again. In column 2 put the requested information opposite the appropriate format heading. Also look for implied needs that are not directly stated.

SPREADSHEET EXAMPLE

Required Format	Required Information
ESCO qualifications	*We are incredibly good*

Step 3. Go back and scour the RFP, read between the lines and add to column 2 the other key things you have found through careful analysis of the proposal request. These details in column 2 will begin to differentiate

your responsiveness to the RFP from the competition.

Technical Capabilities *Auditing*	*Predictive Consistency* *"Our actual savings fall within ± 5% of project estimates 95% of the time."*

Step 4. Go over your notes from on site discussions, walk-throughs of the facility, phone conversations, pre-proposal conference, research, etc. Information may come from surprising sources. If you are proposing to a school system, for example, review recent board minutes to identify what special needs or wishes may have been mentioned. You might even point out how energy savings attained in the project could fund certain expressed needs. One ESCO, for instance, determined that a Florida school system needed a new phone system, added this to the proposed scope of work, and explained how the energy savings would pay for the phones. They got the job.

Do not overlook the operations and maintenance (O&M) personnel as a source of vital information. More than once their voiced special concerns/needs have made a proposal sing. O&M people seldom sign the contracts, but they can voice a critical, "No!" Always keep in mind that they may not have the final say, but their "No" can stop you dead in your tracks. And if you get the job, they are often the difference between success and disaster. Just ask yourself, "How will the O&M people react during project implementation if they have declared loudly that performance contracting won't work?"

Once you have gleaned these special insights into the potential customer's needs, go through and add these thoughts to column 2. It really pays to double check that all you have learned is incorporated in your remarks related to the required information in column 2.

Step 5. Now it's time to drag out those special embellishments that set you apart from the competition. Even better ... if your forte has a perfect fit with the owner's needs, then play it up big. On your spreadsheet, create column 3 for the "Touch of Class" you bring to the table. Determine the appropriate format headings where you should note the special qualities you bring to the project and enter them there. Keep in mind the old line, "If you don't toot your own horn, no one will know you can play."

Modesty at this stage is not particularly helpful.

A little one-upsmanship could be key. If you know what your probable competition will be touting in their proposal on this job, figure ways to one-up them or neutralize them.

Keep in mind that many things your ESCO does in the course of the job can be sold as extras to the owner. For example, you cannot adequately assess potential savings without analyzing the utility rate schedule and the billing. It is not uncommon for school districts, for instance, to be on the wrong rate schedule or not be enjoying the benefits of a particular rider. For your potential client, this rate analysis can be a "freebie." Or, you can use the information as part of your sales approach.

Now, for the *crème de la crème*. Identify one special thing you can add, or do differently that will make what you bring to the table unique and very attractive to the potential client. Put in just enough to get their attention.

Be careful, though, as this can easily be over done. It is a fine line between giving the potential customer enough to win the job and giving away the store. Remember they have the right to share every thing you propose with the other ESCOs, or implement it themselves. A savvy RFP writer is going to state in an RFP that the owner is free to use all ideas submitted without awarding a contract to the party submitting the proposal. If the owner does not include this provision, then all the ideas from all the proposers would be out of bounds and there would be few energy efficiency ideas left to implement.

The proposers have to offer enough to entice, but hold back enough so the implementation requires their expertise to make it happen. Specially developed software programs are often used in this manner.

As a further example, consider an acute care hospital that has been given "First Responder" status in an emergency/security scenario. An ESCO might offer chauffage with supply from a stationary fuel cell. The proposal could point out that the fuel cell would offer a power source independent of the grid. During an emergency that includes a power outage, a fully operational facility would have a lot of appeal. The smart ESCO, however, would not go into details as to how it would fully implement this particular concept. It is enough to just dangle the bait.

There is no question; it is a fine line. But knowing where to draw it wins projects.

Chapter 5

An ESCO's Guide to Measurement & Verification

Measurement and Verification (M&V) refers to the process of identifying, measuring and quantifying energy consumption patterns over a period of time. This process involves the use of monitoring and measurement devices and applies to new construction, and existing buildings and facilities. Measurement and verification methodologies are used in performance based contracts, project commissioning, indoor air quality assessments and for certain project certifications. By establishing the standards and rules for future assessment criteria, the concept of measurement and verification is a key component of energy savings performance contracts. In performance contracts, the performance criterion of a project is often linked to the cash flows associated with the facility improvements.

Measurement and verification can be defined as the set of methodologies that are employed to validate and value changes in energy and water consumption patterns over a specified period of time, which result from an identified intervention or set of energy conservation measures. For energy services companies (ESCOs), measuring energy patterns involves a series of engineering assessments of energy usage and costs. The questions that an ESCO might have include:

1) What process needs to be followed for successful M&V?
2) What M&V options are available?
3) What is the relationship between the M&V methodologies and the costs of M&V?
4) What sort of documentation is needed and how often are reports required?

The answers to these questions vary from project to project. Projects are generally considered financially successful when benefits and savings are achieved. There are risks and financial uncertainties associated with energy savings projects. To reduce financial risks, ESCOs will often guarantee a level of savings that is less than the estimated or predicted quantity. Regardless, performance contracting might be viewed as a risk mitigation tool for customers.

Since ESCOs guarantee project costs savings, there is a shift in financial liability for a majority of unrealized savings or loss of savings due to unsuccessful energy measures. However, when savings goals are not achieved, both the ESCO and their customer may suffer financially. To confound the process, project complexity can increase M&V costs.

Thankfully, technologies and methodologies are available to measure, verify and document changes in utility usage. Tools available to ESCOs include M&V guidelines and protocols that establish standards for primary measurement and verification options, test and measurement approaches, and reporting requirements. Using procedures identified in the guidelines and protocols, the ESCO develops a measurement and verification plan to answer these questions and to serve as a guide as the process unfolds. The final M&V methodologies become formalized as part of the contract with the ESCO's customer.

The purpose of this chapter is to summarize the M&V process from the ESCO's perspective. This process typically involves five primary steps: 1) performing the pre-construction M&V assessment; 2) developing and implementing the M&V Plan; 3) identifying the M&V project baseline; 4) providing a post-implementation report; and 5) providing periodic M&V assessment reports. This chapter clarifies this process and offers the ESCO an overview of the planning aspects of measurement and verification.

What Drives The Need for Measurement & Verification?

Energy conservation projects are often financially justified on the savings that result from their implementation. Performance contracts are viewed as a financial tool to maximize a facility owner's capital by providing facility improvements that have an associated cost reduction guarantee. Prior to standardized measurement and verification procedures, little justification other than an engineer's calculations or a manufacturer's data sheet was used to justify the cost savings resulting from the improvements. While these approaches may be used to predict future utility patterns, justifying projects in this manner failed to validate savings impacts after projects were implemented. In the end, there was the question of whether or not the predicted savings were actually realized. Methodologies were needed to satisfy such skepticism. The concept of measuring and verifying savings is driven by an owner's desire to identify and quantify project savings after an intervention has been performed and before any savings-based payment is made.

In the real world, utility consumption patterns are not static. As shown in Figure 5-1, energy consumption patterns fluctuate due to innumerable variables including weather and environmental conditions, facility design, interior space requirements, facility management procedures, level of maintenance, etc. Estimates of energy usages can vary as a function of the calculation techniques employed, differences in application of technologies, site variables and other differences in operating parameters. This variability drove a need develop certain standardized M&V procedures.

In order to more fully justify longer term capital improvements, demand developed for M&V techniques that more precisely documented changes in energy usage and resulting changes in project cash flows. In addition to savings from energy, there was also a need for industry standard M&V procedures that applied to water and sewer improvements, operations and maintenance savings and savings from other opportunities.

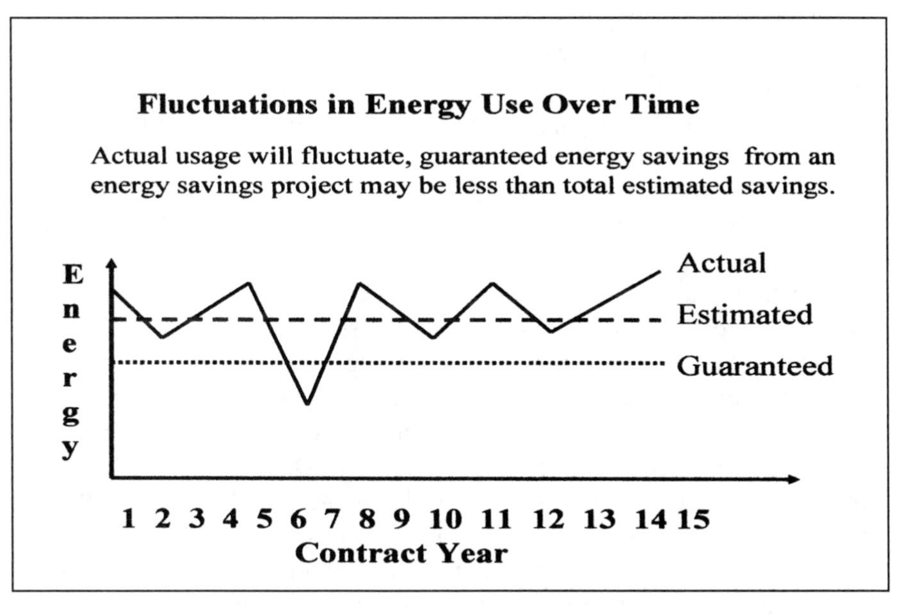

Figure 5-1. Energy Usage Fluctuation

THE USES FOR MEASUREMENT & VERIFICATION

A goal of the M&V process is to periodically validate both project savings and the performance aspects of projects. Criteria are established against which the success of meeting the performance requirements can be measured and assessed. M&V is used to validate commissioning, meet requirements for certain project certifications; e.g., Leadership in Energy and Engineering Design (LEED), validate savings for performance contracts and to verify operational parameters of HVAC equipment. When commissioning involves a guarantee of standard compliance, M&V is useful in providing guidance for baselines. For projects seeking LEED certification, M&V is recommended for lighting systems, wastewater applications and for credit under the energy and atmosphere category [1]. In addition to general requirements for M&V, roughly 40% of LEED projects submitted for certification apply for the credit available in the M&V category. When organizations have energy management programs in place, M&V can provide cumulative benefits over time by helping to keep operations running at maximum efficiency [2].

Perhaps the most widespread use of measurement and verification is for performance based contracts which guarantee savings. Savings for other performance aspects of projects, such as operational and maintenance savings, may also be guaranteed. For performance contracts, the M&V process establishes standards to legitimize project savings, to provide a means of comparing project cash flows to baseline conditions, and to establish the conditions of payment.

Measurement and verification services can be performed by the energy services provider, by the project owner or by a third party that is selected by the mutual agreement of the ESCO and its customer. Third parties may include engineering consultants skilled in M&V procedures and analysis. M&V provides a form of performance assurance, meaning that there is a supportable rational for the guarantees provided in the performance of the equipment and building modifications installed. In order to bring credibility to the process, at least one member of the M&V team should be a Certified Measurement and Verification Professional (CMVP) [3]. This certification program requires that candidates meet specific educational and professional experience criteria and pass a certification exam.

With recent advancements in monitoring and measurement technologies, it is possible for energy engineering professionals to log and record most energy consumption aspects of the energy conservation measures they implement. Examples include the use of data loggers, infrared thermography, metering equipment, monitors to measure liquid and gaseous flows, heat transfer sensors, air balancing equipment, CO_2 measurement devices and temperature and humidity sensors. Remote monitoring capabilities using direct digital controls (DDC), fiber optic networks and wireless communication technologies are also available. As applied monitoring technologies evolve and become accepted by the marketplace, costs for installed monitoring equipment will continue to decline as the capabilities of monitoring technologies improve.

MEASUREMENT & VERIFICATION Protocols

Two primary M&V protocols have been developed in the United States. Early efforts included The North American Energy Measurement & Verification Protocol (NEMVP, 1996) and the Measurement & Verification Guideline for Federal Energy Projects (1996).

The North American Protocol was successively improved and evolved to international stature, becoming the current three volume, International Performance Measurement & Verification Protocol (IPMVP, April 2001). Volume I identifies methodologies and options for determining energy and water savings. Volume II concerns M&V concepts and practices for improving indoor air quality. Finally, Volume III deals with M&V approaches for assessing energy savings from renewable energy sources and in new construction [5]. Each provides details of M&V options and guidelines for verifying energy savings. The methodologies described in the first volume are based on the use of four primary M&V options, which will be explained in detail. The IPMVP is presently the most widely adopted Measurement and Verification standard. A more technical document has been developed by the American Society of Heating Refrigerating and Air Conditioning Engineers. ASHRAE 14 is compatible with IPMVP; and, in fact, Technical Committees of the two groups shared many of the same members.

The Measurement & Verification Guideline for Federal Energy Projects (available at *http://ateam.lbl.gov/mv/*) was updated in 2000, and applies to Federal government projects in the United States [4]. U.S. Federal M&V Guidelines are similar to those provided in the IPMVP, with the primary exception that a more liberal use of stipulations is allowable in the Federal M&V Guideline.

AVAILABLE M&V

The theoretical basis for Measurement and Verification in regard to assessments of resource usage over comparative periods of time can be explained by the following equation:

$$\textbf{Change in Resource Use}_{(adj)} = \Sigma \textbf{ Post-Installation Usages} +/- \Sigma \textbf{ Adjustments} - \Sigma \textbf{ Baseline Usages}$$

Baseline usages represent estimates of "normal" usages prior to implementation of any cost savings improvements. Adjustments are changes in resource use that are not impacted by an intervention and are considered exceptional. The term intervention refers to the implementation of a project that disrupts "normal" energy usage patterns. Post-installation usage refers to resource consumption after the intervention has been performed. Using this formula, negative changes in resource use represent declines in adjusted usage while positive changes represent increases in adjusted usage.

In order to quantify usages, specific measurement technologies and verification methodologies are selected. An understanding of the various M&V methodologies available is critical in order to propose the most appropriate M&V alternative for projects. The most appropriate M&V approach is the "best fit" option that resolves the competing influences of customer needs, the ESCO's M&V capabilities, the methodologies available, the variables to be assessed, the reporting requirements, etc.

In performance contracting applications, the actual measurement & verification approaches utilized are negotiated between the ESCOs and their customers. As a practical matter, ESCOs will initially propose an M&V approach. M&V typically requires that data for numerous measurable variables be collected and analyzed. Measurable variables may include uncontrolled variables such as temperature, relative humidity or pressure measured by a sensing element [5] that will be influenced or managed by the interventions.

Since the IPMVP is the most widely used protocol, its standards will be used as an example. The four measurement and verification options described in the IPMVP (April 2003) are summarized as follows:

Option A: Partially measured retrofit isolation.
Using Option A, standardized engineering calculations (based on product lab testing by the manufacturer) are performed to predict

savings using data from manufacturer's factory testing and a site investigation. Select site measurements are taken to quantify key energy related variables. Variables determined to be out of the ESCO's control can be isolated and stipulated; e.g., stipulating hours of operation for lighting system improvements.

Option B: Retrofit isolation of end use, measured capacity, measured consumption
Option B differs from Option A, as both consumption (usage) and capacity are measured (output). Engineering calculations are performed and retrofit savings are measured by using data from before and after site comparisons; e.g., infrared imaging for a window installation or sub-metering an existing chiller plant.

Option C: Whole meter or main meter approach
Option C involves the use of measurements that are collected by using the main meters. Using available metered utility data or submetering, the project building(s) are assessed and compared to baselined energy usage.

Option D: Whole meter or main meter with calibrated simulation
Option D is in many ways similar to Option C. However, an assessment using calibrated simulation (a computer analysis of all relevant variables) of the resultant savings from the installation of the energy measures is performed. Option D is often used for new construction, additions and major renovations; e.g., LEED-NC certifications for new construction.

The measurement and verification options available in the IPMVP each provide an alternative means of meeting a requirement for verifying savings. Depending on site conditions, each approach will have discrete advantages and disadvantages. For example, in cases where facilities have main meters in place, Option C may be preferred.

Table 5-1 suggests how each of the four options might be used for the measurement and verification of savings for window replacements projects.

Simulations performed for Option D will often involve a computer analysis incorporating the primary design features of the proposed solution. Computer programs used that may be for simulations include BLAST, Carrier E2011, DOE-2.2, eQUEST and Trane Trace among others.

Table 5-1. Applying M&V to Window Replacement

Applying Measurement & Verification Methodologies—Window Replacement

Option A: Partially measured retrofit isolation

- Perform standardized engineering calculations (based on manufacturers lab testing of product) to predict savings using data from manufacturer's factory testing and a site investigation.

Option B: Retrofit isolation of end use, measured capacity, stipulated consumption

- Perform calculations and measure savings from retrofit using data from before and after site comparison (e.g. infrared imaging).

Option C: Whole meter or main meter approach

- Use utility data available or sub-meter the buildings in the project in addition to Option B analysis.

Option D: Whole or main meter with calibrated simulation

- Perform all of the above and assess with calibrated simulation (computer analysis of all relevant variable) the resultant savings from the installation.

ANSO/ASHRAE/IESNA Standard 90.1 (2004), *Energy Standards for Buildings Except Low-Rise Residential Buildings*, provides general requirements for simulations that involve the use of the energy cost budget method, a method applicable for certain M&V applications.

COST OF MEASUREMENT & VERIFICATION

Measurement and verification is a data collection and data analysis process that can be costly. The costs vary as a function of the nature of the guarantee, the methodology selected, types of utilities to be metered, the types of equipment required for monitoring and the labor required to perform the data collection and analysis. In performance contracts, savings might accrue from reduced utility usage, reduced demand costs, reduced transportation costs, or from other sources.

ESCOs or third parties typically charge fees to perform the M&V engineering analysis. In M&V contracts, the fees may be included in the initial project costs, or more typically, charged annually after the periodic assessments. In some contracts, the first two or three years of M&V are included in the contract with the customer having the option to continue M&V for an annual fee. In others, the customer may have the option to discontinue M&V assessments and fees but may be required to release the ESCO from future reporting and guarantee requirements.

To minimize the costs of measuring and monitoring the energy savings, a form of M&V Option A is often used. When higher levels of uncertainty (risk) in quantifying savings are acceptable, Options A or B are often selected. M&V costs for these options can be mitigated somewhat by using statistically significant random samples when there is a large number of similar components involved; e.g. lighting fixture retrofits.

The risks associated with greater uncertainties can be reduced when Options C or D are used. However, as more sophisticated procedures and technologies are employed to quantify energy savings, M&V costs tend to increase. Thus, decreases in uncertainty are purchased with increases in

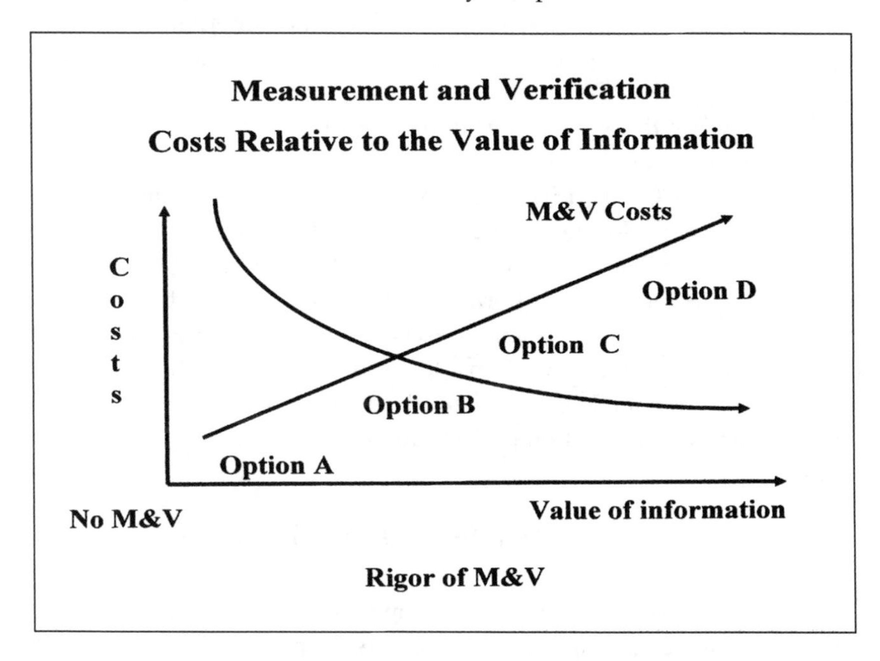

Figure 5-2. Costs Relative to Information Yielded

M&V costs. The added costs are due to the increased use of metering, submetering, digital equipment and more complex monitoring systems (hardware) that aid in capturing information to quantify savings. Such equipment might include sensing devices (temperature sensors, flow meters, pressure switches), communications equipment and programmable controllers.

Options C or D may involve installation of remote monitoring equipment, sophisticated metering devices and connections to existing or proposed digital energy management systems (EMS). More complex technologies often require more data gathering, calibration, maintenance and data management (software) capabilities. However, if data collection systems; e.g., a building automation system that reads metered utilities, are already in place prior to the M&V project, the costs for M&V can be reduced by using the existing systems or adding appropriate metering and monitoring equipment to the existing systems.

As a practical matter, many projects use a single M&V procedure for all implemented measures. However, it is not uncommon for the M&V Plan to specify different M&V options depending on the type of measure being implemented, even within the same facility. For example, a water conservation project for a facility might use Option A while other measures impacting energy usage might use Option C. Alternatively, a lighting project might use Option B while a new building addition necessitated the use of Option D. One motive for mixing options might be to reduce M&V costs when there are opportunities to use less expensive approaches for certain energy efficiency measures.

For the ESCO, the difference in M&V approaches becomes a matter of how their M&V specialists manage the mathematical analysis. While ESCOs may offer all M&V options to their customer, in practice, they tend to specialize in their M&V offerings. Some ESCOs are Option A shops, with a high percentage of their projects using some form of Option A. Others tend to be Option C shops, focusing to a great extent on whole meter analysis. Other ESCOs, especially equipment suppliers primarily involved in new construction projects, tend to use primarily Option D.

The Measurement & Verification Process

Properly managed measurement and verification provides numerous benefits for ESCOs and their customers. Implementing M&V strategies in energy performance contracts is a means of verifying the achievement of energy cost savings guaranteed in the contract over a period of time. Benefits include allocating risks to appropriate parties, reducing uncertainties to acceptable levels, monitoring equipment performance, finding additional savings, improving operations and maintenance savings, verifying that cost savings guarantees are being met, and allowing procedures for future adjustments as needed. A quality M&V Plan may also help secure project financing—and possibly reduce rates.

Measurement and Verification benefits can be maximized by following a few basic procedures. The primary steps in the M&V process typically include:

Step 1: Performing a pre-construction measurement and verification assessment
Step 2: Developing the measurement and ferification plan
Step 3: Establishing a measurement and verification baseline
Step 4: Providing a post-implementation measurement and verification report
Step 5: Providing periodic assessment reports

Each of these steps will be discussed in detail. For purposes of clarity, the five steps are described as sequential yet may not necessarily be performed sequentially. Regardless, Step 1 must occur prior to the intervention. Step 2 provides an action plan which defines subsequent activities. Steps 3, 4 and 5 require reporting to document results. Step 4 occurs only once (shortly after project completion). Step 5 occurs periodically, using data to compare utility usage and costs to the baseline developed in Step 3 [6]. ESCOs typically use an annual M&V assessment for periodic reporting purposes. While there is variability in how ESCOs manage the M&V process, the primary components of all five steps are necessary for successful M&V.

STEP 1: PERFORMING THE PRE-CONSTUCTION ASSESSMENT

During project programming and design, energy and water saving measures are considered and evaluated for implementation. The bundle of energy conservation measures (ECMs) developed for a performance contract can be comprehensive in scope, encompassing many types of facility improvements. In performance contracts, bundling energy and water saving measures is often a means of matching specific project financial goals to customer equipment needs. This creates the opportunity to combine short term payback energy measures with longer term payback measures. The end result is a mid-term payback project that optimizes the financial requirements of the project.

Bundling energy and water conservation measures can cause savings assessments to become more complex, particularly when multiple ECMs impact a single facility. For example, reducing water usage may reduce electrical use by pumps and the energy needed to heat water. Alternatively, reducing lighting system electrical energy usage may increase space heating loads while reducing space cooling loads. In such circumstances, the interactive effects upon energy flows from multiple ECMs must be calculated. The potential for these interactions are often identified during preliminary on-site inspections, and may limit the value of using the retrofit isolation Options A and B.

During the pre-construction M&V assessment, the physical site conditions that impact energy, water and sewage usage are identified and documented. This assessment provides an inventory of energy consuming devices and equipment; e.g., personal computers or vending machines, found on site. The inventory will include not only equipment that may be modified or replaced as part of the project but also other equipment whose energy usage has potential to necessitate an adjustment during future measurement and verification reporting periods.

The pre-construction assessment is codified in a report that becomes a part of the initial project proposal. This report documents existing site conditions, inventories energy and water consuming equipment, and records operating conditions and parameters. The project proposal includes a listing of individual energy measures with identified costs and savings. When savings are discretely identified and separable, measurement and verification for each of the measures can be performed on an itemized basis.

In practice, given the potential interactive effects among ECMs, the savings attributable to the individual measures may be difficult or costly to quantify. In such cases, M&V options that employ partially measured retrofit isolation, whole meter analysis or computer simulations might be the most appropriate.

Typically, building owners are usually more concerned with verifying the total bundled project savings rather than discretely verifying savings from multiple individual measures. For performance contracts, savings from the implemented measures are often aggregated for M&V purposes. The aggregated utility savings are the total savings that accrue from the combined project. As a result, total project savings typically becomes the basis for periodic savings comparisons. Pre-construction M&V assessments focus on collecting current information about the project site and building utility consumption that will be important for these comparisons.

While aggregating the savings from energy and water conservation measures can reduce M&V costs, aggregation has drawbacks. Aggregation limits the ability to attribute savings to specific measures. Aggregation may also increase the difficulty of estimating the impact of future modifications to facilities and equipment. This means that certain savings or costs credited to the interventions during the periodic assessments may be unwarranted.

STEP 2: DEVELOPING THE MEASUREMENT & VERIFICATION PLAN

After assessing site conditions and identifying the measures to be implemented, the project's Measurement and Verification Plan is developed*. The M&V Plan serves as the cornerstone of the energy services guarantee process and is key to the success of the performance assessment. The project's M&V Plan is an agreement that is negotiated between the performance contractor and owner that will be included as part of the final contract.

The purpose of the Measurement and Verification Plan is to identify and codify the procedures, methodologies, measurement devices, standards and processes that will be used to effect the measurement and verification of the program. The M&V Plan establishes the rules by which the M&V evaluation process for a particular project will proceed. While an applicable M&V standard is normally identified in the M&V Plan, the plan itself is a customized document, designed to meet the requirements of a given project. The components of the measurement and verification plan typically include the following:

---

*A template for M&V plans, which has been developed for Federal ESPC projects, has broad applicability. See http://ateam.lbl,gov/mv, *M&V Plan Outline*, November 2004. This web site also offers additional M&V resources.

- List of applicable buildings and facilities.
- List of existing and proposed energy consuming equipment.
- Lists of ECMs that require measurement and verification and also improvements that are not subject to measurement and verification.
- Term of the M&V project and the term that applies to each savings component.
- Project utility usage and cost baseline.
- Requirements for measurement of functional capacity (output).
- Methodologies for calculating costs and savings.
- A listing of applicable standards.
- Product and equipment warranty verification requirements.
- Equipment maintenance requirements and criteria noting responsible parties.
- Schedule for post installation report and post installation site inspections.
- Schedule for required periodic M&V reports (typically reconciled quarterly, semi-annually and/or annually) with descriptions of allowable adjustment criteria.
- Identifies who is responsible for performing M&V and details ESCO and owner responsibilities and obligations.

The M&V Plan will elaborate on each of these issues and include others that are specific to the project. For example, owner's responsibilities might be identified as providing access to the site for the ESCO's M&V specialists, supplying utility bills, alerting the ESCO of changes in facility use or facility modifications; e.g., building additions, providing equipment maintenance, providing updated equipment inventories and being available for periodic M&V reviews. The M&V Plan might include a statement as to how excess savings are to be distributed and how shortfalls are handled. While a few projects use only one measurement and verification methodology, more often, multiple options are used to assess the multitude of technologies used in a performance contract.

For sites with multiple facilities and projects with multiple energy and water savings measure, M&V Option Application Matrices are developed to identify the selected options used for each measure at each site location. Typically, a two-way matrix lists the selected energy, water and maintenance savings measures on the X-axis and the site locations on the Y-axis. Intersecting locations on the matrix identify the applicable Measurement and Verification Option (A, B, C, or D) to be used in Table 5-2.

M&V Plan Option Application Matrix

MEASURES	BLDG A	BLDG B	BLDG C	BLDG D	NEW BLDG
Lighting System Upgrade	A	A	A	A	D
Vending Miser Controls	B	B	B	B	
Replace Exit Signs	A	A	A	A	D
Water Conservation Measures	A	A	A	A	
Window Replacement		C		C	
Install Solar Hot Water Heating System			A	A	
Install Energy Efficient Motors	A	A	A	A	D
Replace Chiller					D
Install Energy Efficient Boiler	C		C	C	D

35

Table 5-2. M&V Option Application Matrix

Criteria for energy usage adjustments are also provided in the M&V plan. In this way, newly applied standards can be incorporated into the architectural and engineering solutions without causing project savings to be penalized. For example, if light levels in occupied areas are below current standards and need to be increased, an energy baseline adjustment may be allowed to compensate for the added electrical energy required to meet the current standard. If an increase in the quantity of outside air supplied to an occupied space was required as a result of mechanical improvements, energy use would likely increase due to the necessity of conditioning a larger volume of air. Adjustments can be made to offset the increased energy use necessary for compliance. In this way, newly applied standards, such as updated building code requirements, can be incorporated into the architectural and engineering solutions without causing project savings to be penalized. In such instances, the energy services provider may be allowed a reasonable credit for any increases in energy use needed to meet the new standards. Other noteworthy adjustments to baseline energy usage that are identified in the M&V plan include additions or deletions of equipment, changes in the plug load energy usage; e.g. soft drink machines or computers, changes in occupancy or occupied hours, and building expansions or additions.

STEP 3: ESTABLISHING THE MEASUREMENT & VERIFICATION BASELINE

Implementing the Measurement and Verification Plan involves data gathering, compilation of data and documenting the results in a series of reports. Energy usage baselines are energy usage profiles (typically for a one year time period) that have been adjusted for extraordinary influences and events. In recent years, projects have also been base-lined against environmental conditions, indoor air quality goals or requirements for code compliance.

Baseline development generally involves regression analysis. The Measurement and Verification Baseline Report provides detailed assessment and analysis of baseyear energy consumption. This document records usage patterns prior to the implementation of any energy and water savings measures. The goal of developing a baseline is to clearly identify pre-installation usages so that future usages can be compared over time. Without a clearly quantified set of base period conditions and utility impacts, future comparative assessments of project associated usages, savings and costs may be invalidated.

Figure 5-3, "Theory of M&V Baseline Development," reveals that actual energy use increases over time above the pre-installation level. Causes for increases in actual energy use might be due to changes such as added equipment, increases in plug load or other unrelated causes. Despite an increase in actual facility energy use, reduced energy use is identified by the M&V process as the difference between adjusted usage and actual energy use. It is also possible for a decrease in base-lined energy usage to occur. Developing a facility energy usage baseline for performance based projects also involves the following:

- Defining a time period (typically a modified 12 month baseyear) for the facility's historical baseline usage;
- Selecting a period for project implementation, during which energy usage may or may not be evaluated;
- Defining a date for the beginning of the performance assessment period.

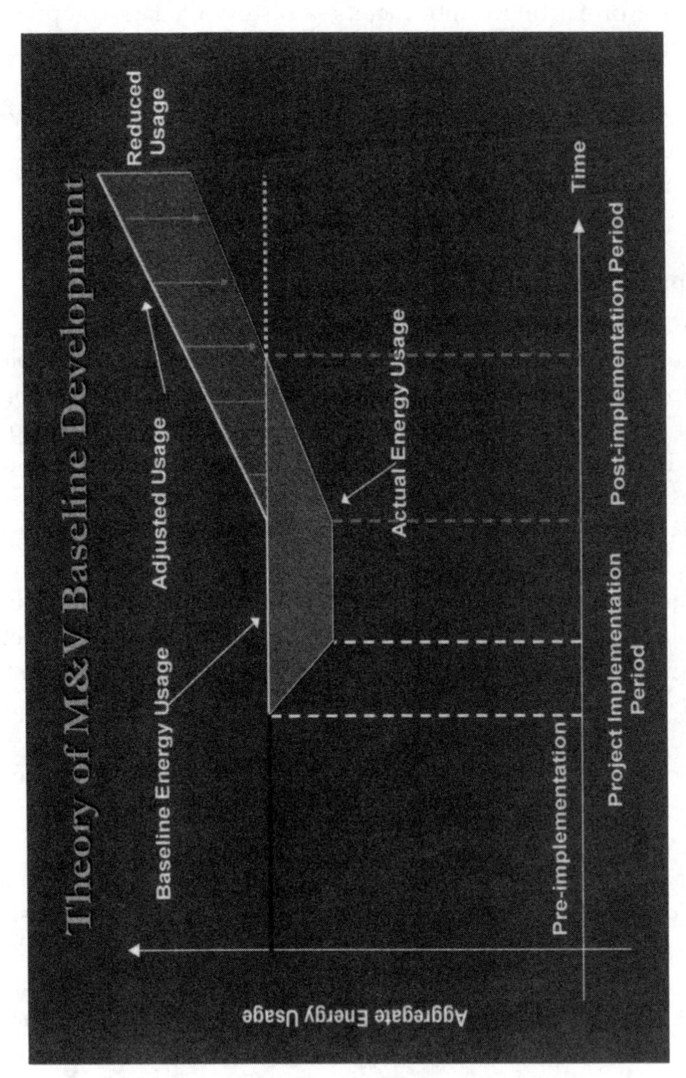

Figure 5-3. Theory of M&V Baseline Development

The identified baseyear may be the result of utility bill analysis or calculations based on observations and known events that resulted in energy and water usage over a period of time.

Information compiled for the energy usage baseyear typically includes a list of existing utility meters, a history of data compiled from utility bills, interval demand billing data, utility rate sheets and any site information that would assist in rationalizing energy use patterns. When whole building assessment options (Options C or D) are employed, all building equipment and operational conditions require documentation and are included in the final report.

STEP 4: PROVIDING A POST-IMPLEMENTATION REPORT

After the physical implementation of the project, a post-implementation report is developed by the ESCO and provided. This is not to be confused with a "punch list" that might provide a list of discrepancies that require correction prior to project acceptance. The primary purpose of the post-implementation report is to document the differences between what was initially intended to have been installed and what was actually installed. It focuses on those changes that may result in a change in energy use and usually requires a site visit by the M&V team. As a result, the post-implementation report identifies and resolves discrepancies between the design intent of the project scope of work and the actual implemented project scope. This requires a review of the final proposal and a comparison of scope with as-built conditions.

More than simply verifying installation of equipment, the post-implementation report identifies all changes in project scope that materially impact project utility savings and costs and quantifies their impacts. Commissioning activities, equipment calibration and performance deviations are documented. From this information, it is possible to reassess the established post-implementation energy usage accordingly. In addition, the post-installation report often provides an inventory of any primary energy consuming equipment in place that was not impacted by site improvements. Electrical loads, equipment sizes, existing fixture counts, plug load survey and facility operating parameters may also be updated if site conditions have changed after the pre-construction assessment was performed.

One example is the identification of differences between light fixtures lamps proposed (counts and energy use) and those actually installed. If the project involved installation of light fixtures and retrofitting a building's lighting system, this report would identify all proposed fixture modifications (providing fixture types, quantities and lamp wattages), indicate the fixtures that were actually installed and identify any differences. Changes in output might be assessed by comparing changes in light levels. Another example might be a change in chiller specification due to the installation of a more efficient chiller than was originally specified.

While project commissioning often occurs during equipment installation, continuing through the first year after project completion, M&V is typically a longer term effort. Regardless, Step 4 may be replicated to some extent by facility commissioning activities. It is important that care be taken to ensure that the efforts are not duplicated—and paid for twice.

STEP 5: PROVIDING PERIODIC MEASUREMENT & VERIFICATION REPORTS

Periodic reports are developed and provided as the M&V Plan is implemented over time. The periodic reports compare energy use during the reporting cycle to baseyear energy use, most typically on an annual basis. However, quarterly and semi-annual reporting periods are not uncommon, particularly in the first year of a verification term. These reports tend to be comprehensive, utilizing software analysis tools to incorporate the influence of key variables.

The periodic reports will document the performance of the project as compared to the stated goals by quantitatively tabulating comparative savings and costs. Actual energy use (in units of fuels used) is compared to baseyear energy use as defined in Step 4. Of concern to the ESCO and their customer, is the relationship between the actual savings resulting from the project and the predicted savings, which were guaranteed.

How the distribution of shortfalls is handled depends on how the performance contract distributes project risk. Typically, savings shortfalls, are distributed among stakeholders as defined in the M&V Plan or the performance contract. In most cases, savings for performance-based projects will meet or exceed projections. Current energy usage is compared to initial period baseyear energy use in order to estimate the savings resulting from the project for the reporting period. Regardless, the periodic reports often reflect adjustments that are necessary and available due to changing, and perhaps unforeseen, conditions that may have impacted energy usage during the reporting cycle. Potential adjustments may be allowable as a result of weather conditions, changes in electrical demand, changes in use of activities, building additions, or installation of additional energy consuming equipment, if permitted by the Measurement and Verification Plan.

Procedures for adjustments are identified in the M&V Plan. For any given project, additional adjustments may be provided and customized as the special circumstances of the project dictate. Adjustments can be managed with the use of spreadsheets or computer software programs such as FASER™ and METRIX™.

SUMMARY

The Measurement and Verification Plan is the cornerstone of the M&V process, since it identifies the standards against which the resulting project savings are compared. In this chapter, the types of information needed to develop a M&V Plan were identified. Components of the process include developing an M&V baseline, selecting from the available M&V options, designing an M&V Plan, implementing the plan and reporting results.

M&V information offered in this chapter provides management guidance to an ESCO or a potential performance contracting customer. The chapter is not intended as comprehensive guidance for an M&V specialist or practitioner who needs a working knowledge of the M&V protocols that apply in practice.

In the future, it is likely that Measurement and Verification will prove to be useful for a broader range of technical applications. Some M&V approaches will become routine. As digital technologies become more widely used, it is likely that the cost of data accumulation and management will decline as the accuracy of measurements improves. Computer software for measurement and verification will be more widely available and user friendly. In addition, M&V reporting methodologies are likely to be refined and become more standardized.

References

[1] U.S. Green Building Council. Green Building Rating System for New Construction and Major Renovations Version 2.1 Reference Guide.

May 2003. pgs. 73, 103, 173-181.

[2] ION Wire, Improving Energy Efficiency with Measurement and Verification, Issue 19, December 2002.

[3] Association of Energy Engineers, M&V Professionals Certification Board, Lilburn Georgia.

[4] U.S. Department of Energy and Office of Energy Efficiency and Renewable Energy. *M&V Guidelines: Measurement and Verification for Federal Energy Projects*, Version 2.2. DOE/GO-102000-0960. September 2000.

[5] IPMVP New Construction. *International Performance Measurement & Verification Protocol, Vols. 1, 2 & 3*. April 2003. See http:/www.ipmvp.org.

[6] Petrocelly, K.L. and Thumann. A. (1999). Facilities Evaluation Handbook. Fairmont Press. p. 257.

[7] Roosa, S.A. AEE-ETE Conference, Energy Efficiency, Energy Markets and Environmental Protection in the New Millennium, Sopron, Hungary, 15 June 2001, *"The Energy Engineer's Guide to Performance Contracting."*

Chapter 6

Financing Energy Efficiency

No matter which way we turn in performance contracting, financing confronts us as a critical issue. Customers are driven to seek performance contracts because of the financing. Energy service companies (ESCOs) manage and cushion their risks through the financial structure of a project. Securing the most favorable financing terms affects the quality of a project for both the customer and the ESCO. Clearly, financing is a major element in performance contracting and ESCO operations.

Before we turn to financing specific energy efficiency projects, a couple of thoughts on broader financing strategies for the ESCO should be addressed. Financing is so basic to performance contracting that every ESCO needs someone qualified to oversee the company's financing strategies and to analyze the financial component of every project during preliminary assessment.

One aspect of the broader financing strategy should be constant oversight of existing debt. Whenever interest rates start to climb, a quick review of any higher cost debt should be made and refinancing should be considered. The financial officer should also consider extending the terms of the ESCO's bank lines and should keep watch on ways to reduce working capital requirements.

One final observation on the qualities of the financial officer: he or she needs to be multi-dimensional with an understanding of total business operations, individual project financing issues, and factors that influence financing, including finance models and project risks.

Identifying risks associated with the savings guarantee and figuring ways to manage and/or mitigate those risks are discussed in chapters 10 and 11. In this chapter, we are concerned about the money side of that risk management. Proving to financiers that the risks have been identified and effectively managed is crucial to project financing. Whether real or perceived, bankers peg interest rates on what they deem to be the level of risks associated with the project. Presenting a project in the most favorable way, therefore, not only gets the funding, but gets the most favorable terms.

Bottom line: ESCOs, who get the best terms can deliver better projects, have more satisfied customers, become more competitive in the marketplace—and make more money!

When we follow the money trail, it becomes evident that getting a project financed should be a shared effort between the ESCO and the customer. Part of the ESCO responsibility is to be sure the customer understands that this cooperation is important to the bank and will often result in more project for the amount invested. Unlike our friends in the illustration, all arrangements should be clear and transparent.

In the process, however, it should be acknowledged that the perspectives are different. It is the ESCO's responsibility to put together a bankable project. The ESCO typically arranges the financing. Its reputation and history often add surety, which offers financiers added confidence that they will get their investment back in a timely manner. Since the customer usually incurs the debt, owners need to know exactly what financing options are available and the implications of each option. Chapter 11 will underscore the key role the customer can play in keeping risks under control.

First, we will look at what constitutes a bankable project from the ESCO perspective. Then, the types of financing available to owners will be explored.

ESCOs, who have been in this business for a few years, remember knocking on the financial doors until their knuckles were bloody. Today, the financiers knock on the ESCO doors... if, and it's a big **IF**, ESCOs can put together *bankable* projects. ESCO financing has evolved to the point that should an ESCO not be able to get a project financed in the US, it's time to rethink the project.

CREATING BANKABLE PROJECTS

What is a "bankable" project? Simply put, it is a clearly documented economically viable project.

Building a bankable project starts with sorting out the pieces that

make a project economically viable. The first step is to examine the key components and make sure each aspect and any associated risks are properly assessed and a plan put forth that effectively shows how those risks will be managed. Every measure, each component, typically carries a risk factor, and the management/mitigation of each risk factor is apt to carry a price tag. An effective ESCO knows how to assess the components and how to package them into a project that can be financed.

THE CUSTOMER

Pre-qualifying customers is an art. The critical factors for the ESCO are developing the selection criteria, asking the right questions, and *learning to walk away* when a "lucrative" project doesn't match those criteria.

Ironically, one of the major drivers of performance contracting is the owner's need for financing; so it seems like a dichotomy that a primary pre-qualification for a customer that needs financing is to be creditworthy. But a customer can be cash poor and creditworthy at the same time. In fact, a potential customer who is creditworthy and cash poor is an especially promising candidate for a performance contract. A school district, for example, is typically creditworthy and legally backed by the state, but its revenue stream is often sparse.

Most ESCOs have an understanding with a financial house (or houses) as to what constitutes acceptable credit standing. Some even have prescribed forms for the ESCO's sales people to fill out; so all the pertinent information is acquired and presented in a routine fashion. The credit check at this stage is like most others. Financiers want the information that can reasonably assure them that the loan will be paid back.

The range of information a financial house will need regarding a potential customer typically includes:

- the type of transaction proposed, e.g., equipment title provisions, purchase options, payment terms, and the performance contracting financing model to be used;
- the organization's tax status;
- longevity of the customer's organization; ownership;
- its business prospects;
- evidence that the customer can keep the savings, which will provide the all important revenue stream from which the payments will be

made as well as an incentive for the owner to participate;
- financial condition with three years of complete and current financial statements; i.e., bond rating, 10K, or audited financial statement; and
- preliminary project calculations.

The critical financial information needs to be adequately documented. No matter how charming, persuasive and attractive a potential customer may be, the financials must be in print—and signed. In their zeal to make a sale, sales people are sometimes tempted to take the customer's word for its credit standing. But the financier won't. ESCOs can become blinded by the "savings opportunity" and spend a lot of money developing a project based on false financial assurances only to eventually learn the owner cannot meet the necessary financial criteria. This is an expensive lesson for ESCOs. Unbelievably, some need to learn it more than once.

In addition to the customer's creditworthiness, financiers are more inclined to loan money when larger ESCOs are involved. Their size and track record often offer the surety needed to provide lower interest rates. Smaller firms, however, need not be discouraged by this apparent market advantage; for the small firm can typically achieve the same status by getting performance bonds or insurance to cover the savings guarantees. Furthermore, even with these added costs and higher interest rates, the small firm can still compete with the margins charged, for example, by an ESCO affiliated with a manufacturer.

The above concerns relate to the financial pre-qualification of the customer. Once the ESCO is satisfied with the customer's creditworthiness, consideration can be given to other criteria which will be used to weigh the customer's partnership quality, including the administrative commitment to the project, the attitudes and abilities of the operations and maintenance people, etc. These "people factors" and other critical concerns are generally folded into a scoping audit that assesses project potential. The scoping audit is little more than a walk-through audit with a very educated eye. The purpose is to be sure that further pre-qualification and marketing efforts are warranted.

Start-up ESCOs, or "WISHCOs"—a name too often earned—are inclined to overlook the risks associated with the people factor. Since it is vital to project success, it cannot be over-emphasized.

Once the other pre-qualification criteria have been met and the potential customer has accepted the concept, then a full feasibility study is needed (investment grade audit). Before the ESCO incurs the expense of a premium quality energy analysis, an agreement to cover the costs of the audit if the project does not go forward is increasingly used to protect the ESCO's investment. The content of this planning agreement is discussed later in the contracts chapter.

ENERGY AUDIT QUALITY

A standard energy audit with its "snap shot" of current conditions is not good enough for performance contracting. These audits typically assume present conditions will prevail for the life of a project. When an ESCO bets money on predicted *future* savings, these assumptions must be tested through a careful risk assessment procedure.

Only an *investment grade audit* (IGA) that adds specific risk appraisals to the standard name plate/run calculations will meet performance contracting needs. In recent years, energy engineers have learned to look at facility and mechanical conditions and determine the ability of the remaining equipment and energy consuming subsystems to accept the recommended measures. An investment grade audit (IGA) goes beyond these engineering skills and requires the art of assessing people including; the level of commitment of the management to the project, the extent to which the occupants are informed and supportive as well as the O&M staff's abilities, manpower depth and *attitude*.

For the reader, who must perform, oversee or assess the quality of an audit, it is strongly recommended that he or she read *Investment Grade Energy Audits: Making Smart Energy Decisions*, which is available from The Fairmont Press.

A key aspect of a quality IGA is a carefully detailed baseyear with the average energy consumed over several years *and the operating conditions*, which caused that consumption. (See Chapter 11 in the Risk Management section for a full discussion of baseyear issues.)

The ESCO that consistently delivers a quality IGA, which in turn accurately predicts potential savings, builds a track record that financiers find very heart warming. A good IGA is at the heart of a bankable project. When the total project plan is wrapped around a quality IGA and delivered by an ESCO, who can back its predictions with a solid history of successful projects, financiers smile.

EQUIPMENT SELECTION AND INSTALLATION

Predictive consistency, the hallmark of quality audits, rests on knowing what works. And what doesn't! To support a guarantee, ESCOs must have considerable control over the equipment specifications and the selection of the installation subcontractors. Generally, this control manifests itself in order of preference from the ESCO's point of view in (1) working as a general contractor or construction manager, which supplies all the equipment and installation; (2) having primary responsibility for developing the specs in cooperation with the owner and making the final equipment selection; and (3) preparing specs in cooperation with the owner and identifying acceptable bidders for the owner's final selection. For the owner these options offer progressively more control and increasingly transparent costing. The more control an owner exerts, however, the more risk the ESCO assumes, the lower the project economic viability becomes, AND the project bankability drops accordingly.

A financier's due diligence carefully assesses the ESCO's ability to make good on its guarantee and to control the variables that threaten the savings and the guarantee. As always, money follows risk. It's worth repeating: interest rates are directly related to the project risks as perceived by the financier.

For both parties, the predicted benefits must outweigh the expected risks or the project is not bankable. It follows that the control exercised by the owner directly affects the project benefits—inversely. The owner control level translates directly into ESCO risks, project vulnerability and viability—and higher interest rates. Money that goes to pay interest is not available to buy services and equipment, which produce the savings. So owner control has a price tag that has a negative impact on the project.

PROJECT MANAGEMENT

One of the great appeals of performance contracting is the extent to which the ESCO's fee, and profit, rides on the project's success. The truly successful ESCOs know the project is only beginning once the construction/installation/commissioning is done. There are three key components to managing a project, which are closely related to its success. A good bankable project presentation pays close attention to each one; and so does the knowledgeable financier. They are summarized here in terms of presenting a bankable project with a more complete discussion of project management in Chapter 8 and communications in Chapter 9.

1. **A planned effective partnership**. This critical aspect is the most obvious and most frequently ignored. It rests on a carefully orchestrated communication strategy where:

    a) problems are aired, not hidden, and resolved collectively;
    b) successes and the means of communicating them to the customer's internal and external publics are developed in concert;
    c) day-to-day incidents are shared and resolved with a sense of camaraderie;
    d) the ESCO's Project Manager identifies problems and offers business solutions as an adjunct to the customer's operation; and
    e) the communications strategies are reviewed and enhanced as needed for the life of the project.

2. **Maintenance** (and operations to a lesser extent.) Maintenance and operations must be carefully planned and executed in a routine fashion appropriate to the installed equipment. This maintenance may be performed by the ESCO, its trained representative, or the owner's personnel. A checklist and routine policing are needed in all cases.

    An evaluation of the federal energy grants program for schools and hospitals underscores the critical need for effective operations and maintenance (O&M). A study for the U.S. Department of Energy revealed that in an effective energy management program up to 80 percent of the energy savings are due to energy efficient O&M practices; **not the hardware**. Without a good O&M program, guarantees are very difficult to achieve.

    Owners frequently want to keep the maintenance responsibilities for a variety of reasons, including a sense of control, personnel needs, or union issues. If the primary reason is a perceived economic advantage, owners should be aware that ESCOs view reliance on owner maintenance as a significant risk and will financially structure their projects to protect themselves against this risk. In the long run, an owner is apt to budget more for maintenance than it would have if the function had been outsourced to the ESCO. In the process, the owner receives a smaller project.

    In some instances, a computer-based maintenance management

software (CMMS) program may offset the ESCO risk sufficiently to make owner maintenance an economically viable option. The financials for both scenarios should be worked through and compared before owners decide to "save money" by doing their own maintenance.

No matter which procedure is used to achieve a quality maintenance program, a solid CMMS can be an extremely valuable tool. It can serve as a checklist and a record as well as providing work orders, etc. necessary to monitoring a program and assessing the situation should something go wrong.

3. **The Project Manager**. It is impossible to overstate the key role a good Project Manager plays in achieving energy savings and in fostering a strong sense of project partnership. From the start, he or she should help with the risk assessment, help determine customer needs, document needed O&M staff training and personnel augmentation, merge ESCO and owner staff into one team, and become the link between the ESCO and owner management for the life of the project.

MEASUREMENT AND SAVINGS VERIFICATION

When money changes hands based on the level of savings achieved, all parties should be comfortable with how the achieved savings are measured, verified, and correctly attributed to those who performed the work. This issue, very ably addressed by Dr. Roosa in the previous chapter, has become a "hot button" in the industry and is in great danger of being over-played. Under the financiers' general guidance, the ESCO and owner should jointly decide on the level of verification and attribution necessary. It is basically a case of cost vs. accuracy, and it is possible to reach the point of diminishing returns rather quickly. With measurement and verification (M&V), it's easy for the tail to start wagging the dog… with the savings verification burden becoming so great that a measure is no longer economically viable.

The financier wants some sense that the project benefits are measurable and they are measured through accepted protocols. Too often verification procedures are basically passive, a negative drain on the cash flow, and investors are not interested in funding a gold plated M&V approach that offers little or no return on investment.

In the final analysis, a bankable project is one you, as an individual, would want to invest in if someone else were doing it. If we step back a few steps and view it as a banker would; then, an economically viable, bankable project is simply one which demonstrates good business sense.

THINK MONEY

When the various assessment are complete and the pieces assembled, it's time to think about the presentation to a financier. To do that, it helps to think how the banker will view a project.

It is not as mysterious as some believe. Taking a proposal for funding to a banker or financier is just a variation on a theme. It's another sales job.

A couple rules of the road might help.

1. Know yourself, your company and your project before you approach an investor.
   - Be aware of your strengths and weaknesses.
   - Build on your strengths.
   - Acknowledge your weaknesses and how you intend to augment/compensate for them
   - Keep both aspects in perspective.
   - Know why someone should invest in your project and in you.

2. Be aware that every strategy has financial implications and risk parameters. During proposal preparation, the task is to determine:
   - Is the strategy financially sound?
   - Have the risks been adequately addressed?
   - How will investors react to what you propose?

3. Remember bankers rent money. That's the business they are in. You are the customer. The days of going hat-in-hand for a loan are past. *They need a quality project to invest in every bit as much as you need the funds.*

4. We need to go beyond finding the project financing to finding the best deal on the money. Terms vary and quality projects should get the very best terms.

5. But the generalizations stop here. We need to avoid over generalizing about investors. Your challenge is to find financiers who want to invest in what your company can deliver. If they understand your business, it will temper their perception of risk and you will get better terms.

While writing a proposal, it is well to remember the difference between the bankers focus on "return on investment" and the risks ESCOs regularly work with. Returns are relatively easy to quantify, but risks typically aren't. It helps to show investors the risk management strategy you have and how you intend to execute that strategy. The approach is to acknowledge the risks exist, but to take complex issues and simplify them for investor consumption. In short, show them you know what you are doing, but don't belabor it.

A story might help make the point. A little boy went to his mother and said, "I have a question about turtles." The mother responded, "Then, why don't you ask your father, he is a marine biologist." To which the little boy responded, "I don't want to know that much about turtles." Bankers don't want to know that much about turtles either; they only want to know that they will get their money back with interest.

The Owner's Perspective

The first step for an owner in achieving the most effective financing is to get an ESCO that can deliver a bankable project. The ESCO's track record and its bank relationships can tell the owner a lot about that.

Roughly 95 percent of the performance contracts in the United States are currently structured for guaranteed savings with the owner typically accepting the debt through third-party financing (TPF). TPF is especially attractive if the owner qualifies for tax exempt financing. Since the debt will be on the customer's books, owners have some important choices to make regarding that financing.

1. For a *tax-exempt* organization, the project costs can be reduced by thousands of dollars if tax-exempt financing is used. Notice the words used were "tax exempt;" not "if you don't pay sales tax." *Tax-exempt* is clearly defined by the Internal Revenue Code, in 103a, as an organization that can levy taxes, raise a police force and/or con-

demn property. A school district in Florida, for example, is not tax-exempt; for it does not fit the definition. It is possible, however, for that district to ride on its county's tax exempt status.

2. Leases are often found to be an attractive way to finance performance contracts and are available predominantly in two forms: operating leases and capital leases.

   If debt ceilings or greater indebtedness is a problem, an operating lease, which is off balance sheet, can be very attractive. But the qualifications for an operating lease are pretty narrow; so a certified public accountant needs to be consulted PRIOR to the agreement.

   The majority of energy equipment leases are capital leases. If a lease meets any of the following criteria, it is considered a capital lease:

   - the lease term meets or exceeds 75 percent of the equipment's economic life;
   - the purchase option is less than fair market value;
   - ownership of the equipment is transferred to the customer (lessee) by the end of the lease term; or
   - the present value of the lease payments is equal to 90 percent or more of the fair market value of the equipment.

   Conversely, if a leasing arrangement meets any of the above criteria, it cannot be an operating lease.

   Leases work very effectively with guaranteed savings programs. Articles in the popular press too often imply that leasing is available only with shared savings. Not so. Shared savings is only one type of performance contract. Any performance contract can be structured to use lease financing. The owners' choices, therefore, include whether or not to use leasing and, if they do, what type of lease should be employed.

3. Leasing choices may be influenced by the type of performance contract financing model selected. One option, of course, is to let the ESCO carry the credit risk, but this is surprisingly costly to the owner. When the financing is carried by the ESCO, it usually uses a shared savings approach. Shared savings is defined as a performance contract where the percentage split in the energy *cost* savings

is predetermined and the ESCO typically carries equipment ownership until the end of the contract.

Shared savings is typically not the best option for the ESCO or the owner. The cost of money is higher; so less of the investment goes into equipment and services. Since the deal rests on sharing *cost* savings, it bets on the future price of energy. Risky business, so the money costs more. The ESCO carries both the credit risk and the performance risk; so they get more money to cover those risks.

Financial houses may like shared savings financing since the collected interest is higher, but generally it is not in the ESCO's, or the owner's, best interest. ESCOs, who survived the shared savings era of the late 1970s and early 1980s, are quick to point out another major drawback of this approach: the ESCO gets too much money tied up in financing early projects. Soon the ESCO becomes too highly leveraged to take on any more debt.

To overcome some of the leveraging problem, energy project financiers have stepped in with their capital to form single purpose vehicles (SPV) or special purpose entities (SPE), which help to free up the ESCOs to do more projects.

With established M&V protocols, shared savings has become a little more attractive, particularly to owners who need off balance sheet financing.

Other financing options may be presented to an owner. Unless the owner or someone on the staff is very comfortable with such concerns, retaining the services of a financial consultant is advisable ... and may be the cheapest option.

THE BUY-IN; THE BUY-DOWN

As a final cornerstone to this financing business, owners should not overlook the value of taking an equity position in the project. It is a way to get non-energy related projects incorporated, and/or reduce ESCO and financier risks. A little owner equity can be a powerful leveraging force and make a bigger project possible.

The bottom line for owners seeking to finance energy efficiency is: ask your banker. Find out what the men and women with the money need. Then use their guidance to develop a project. The financier's due diligence, in the end, is the ESCO's and the owner's best guarantee that they have a doable project. To underscore a point made earlier, if an energy efficiency project can't get financed in today's U.S. market, the first step is to rethink the project.

A Primer on Financing

There are a number of financing mechanisms available. They differ based primarily on the type of owner, the purposes for which the financing can be used, and in the legal steps required to effect them. Some are only available to government agencies.

Everyone is familiar with conventional loans available from commercial banks. The other major types of financing vehicles are discussed below. Some do not lend themselves as easily to performance contracting and are cited here mostly to serve as a basis of comparison.

FINANCING MECHANISMS

General Obligation Bonds

Definition: GOs are bonds secured by a pledge of a government agency's full faith, credit and taxing power.

General obligation bonds are payable from *ad valorem* property taxes and require voter authorization. State laws stipulate the conditions to be met. For example, they cannot be issued in California without a 2/3 vote of a municipality's constituency. They are considered the most creditworthy by bond investors; and, therefore, are the least expensive form of financing for issuers. General obligation bonds can only be used to finance acquisition and improvement of real property, as prescribed under the respective state law.

Special Assessment and Mello-Roos Bonds

Definition: Bonds issued to fund projects conferring a benefit on a defined group of properties. The bonds are payable from assessments imposed upon the properties, or from special taxes levied upon the properties which receive the benefit.

Special assessment financings are generally used for infrastructure projects; e.g., roads and sewers, while Mello-Roos bonds fund facilities and services, such as libraries and library services. These types of bonds have been, and continue to be, very controversial in the eyes of the general public.

Revenue Bonds

Definition: Bonds secured by a specified source of revenue or revenue stream.

Revenue bonds have numerous uses. Bonds for water, hospitals, airports, etc. are all examples of revenue bonds, where the revenue from a specific source; e.g., airport, water enterprise, hospital is pledged to repay the bonds. Sometimes a third party is established to collect revenues and to administer the promised repayment for a fee.

Lease-Based Financing

Definition: Financing in which the fundamental legal structure is a lease. These include Certificates of Participation, Lease Revenue Bonds, and privately placed municipal leases.

Lease-based financing differs from debt financing primarily from a legal perspective. The obligation to make debt payments is unconditional. Lease payments, on the other hand, are conditional: they need only be made if the lessee has full use and possession of the asset being leased. Restrictions on issuing debt vary by state and may impose significant conditions. These typically do not apply to leases. All lease-based financings share an underlying structure, described below.

Lease financing is the most common type of local government financing in most states. Generally, the lessee is an owner with a project to fund; e.g., a municipality. From a legal perspective, the lessee undertakes the project; i.e., buys the equipment or makes capital improvements, on behalf of the lessor. The lessor leases the project to the owner, which makes regular lease payments. When the term of the lease is over, the owner purchases the project for a nominal sum, often a token dollar, from the lessor.

Investors fund the lease made by the lessor, in exchange for which they receive the lease payments made by the lessee. This may be done through certificates of participation, lease revenue bonds or the lease document itself. The money investors pay for these instruments goes to a lease administrator (for simple municipal leases) or an underwriter, which deposits it (less the underwriter's fee) with the trustee bank. The lease administrator or underwriter, in turn, makes the funds nominally available to the lessor but, in fact, makes them available to the lessee for its project.

As mentioned above, the lessee is not required to make lease payments until, and unless, it has full use and possession of the project. When the lessee has completed the project and there is something in place to lease, the owner begins to make scheduled lease payments. The lessee deposits its lease payments with the trustee bank, which makes the required interest and principal payments to the investors.

Special provisions are offered should the leased equipment or building be damaged. Under such conditions, the lessee may stop payments, until the project is repaired or replaced. The legal documents require that the repair be made as quickly as possible, so that investors wait as short a time as possible for repayment to commence. Because of this abatement risk, lease-based financing carries a higher interest rate than other types of financing.

There are a number of lease-based financing vehicles:

Municipal Leases

This term is often applied to leases even when a municipality is not involved. This may be nomenclature for tax-exempt entity and is frequently used as a legal definition. It is usually, a simple lease, which is funded by one investor, typically a bank or credit company. The bank funds the lease, and the lessee makes the lease payments to the bank or credit company.

Master Leases

This "umbrella" lease is a variant of the municipal lease with general terms and conditions. As the lessee makes individual purchases or begins individual projects, leases or lease schedules are funded and appended to the master lease agreement. If a performance contract is being done in phases; i.e. 10 buildings at a time on a 60 building campus, a master lease may be used to fund each successive phase. The master lease also serves well when the timing or amount of funds to be needed are not yet known.

Certificates of Participation (COP)

This mechanism allows investors to purchase certificates, which offer evidence of their participation, and enables them to participate in the stream of lease payments being made by the lessee to the lessor. Certificates of participation have much higher costs of issuance than municipal leases, but carry lower interest rates. They are well suited for larger, longer-term projects. In performance contracting, the COP approach may be used as pool financing, which can fund several projects. For the investors, this approach spreads the risk over several projects; thus, diminishing the risks associated with just one project.

Lease Revenue Bonds

Similar to a Certificate of Participation, except that instead of a corporation serving as lessor, one government agency acts as lessor while the jurisdiction needing funding serves as lessee. In these cases, a lessor government issues the bonds, enters into a lease with the lessee jurisdiction, and the lease revenues are pledged as repayment of the bonds.

LEASE AMOUNT

The lease amount begins with the project cost, but it doesn't end there. In general, the following are added to that cost to arrive at the final lease size:

Capitalized Interest

The amount of interest that becomes due during the acquisition or construction period. Sometimes referred to as "interim construction financing" in performance contracting deals. Because the lessee cannot be compelled to make lease payments until it has full use and possession of the project, investors are concerned about being paid during the acquisition or construction period. Investors, therefore, require that the interest amount be "capitalized," or borrowed, through the lease, and set aside to be used to make interest payments during that period. The longer the construction/acquisition period, the more capitalized interest is needed. For example, the interim construction interest on a $4 million performance contract project may be $200,000; so the amount financed to include the capitalized interest would be $4.2 million.

Occasionally, the lessee uses internal funds during construction to avoid this interest cost.

Reserve Fund

An additional amount (usually one year's interest and principal payments) added to the lease amount and deposited with the trustee bank. This fund is used to make interest and principal payments to investors if the lessee is late or fails to make its lease payment. A reserve fund is often required for Certificates of Participation and lease Revenue Bonds, but is usually not considered necessary with municipal leases.

Cost of Issuance

Costs of attorneys, financial advisors, consultants, and incidentals, are usually funded though the lease.

When all is said and done, a $1,500,000 lease may make only $1,280,000 available in project costs with capitalized interest, reserve fund, and costs of issuance all taking their toll.

It should also be noted that at least one ESCO, which brags about not sharing the savings, takes its entire fee from projects similar in process to using the Costs of Issuance. Owners should be aware that this ESCO is getting its fee up front from the financier. This ESCO enjoys reduced risks, avoids the depreciated value of the money, and the owner pays the financing for this fee for the life of the project. While promoted as great value to the owner, the advantage all belongs to the ESCO.

The above primer gives a flavor of the options an owner can consider. The goal is to raise awareness as to the range of financing considerations. It is not sufficient information for an owner to decide all financing issues related to a given project. Unless the owner has personnel on staff comfortable with all aspects of financing, consultation with a CPA or the owner's banker is an excellent precaution. The due diligence of the project financier will benefit an owner, but it will not guarantee that the selected financing scheme is the mechanism that would best serve the owner's needs.

Chapter 7

Quality Contracts

Large, cumbersome contracts serve only one purpose: *increasing lawyers' fees*. Sharp, succinct, easy-to-read contracts can, and should, be the goal of everyone involved. Buyer resistance to long contracts is well documented; so ESCOs should do all they can to keep them short. Good solid ESCOs have managed to pare down their contracts to manageable and readable documents—as short as ten pages.

A contract has basically two functions; to manage risk and establish how the project will be managed. An important management function of any contract is to establish responsibilities—and to answer that all important question when things go wrong: Who's to blame?

Owners and managers should never lose sight of the fact that energy service agreements (ESAs) **are negotiable**. Some ESCOs hand out contracts like they are cast in bronze and the word processor is yet to be invented. Any firm that comes in with a "take-it-or-leave-it" contract and *attitude* is not a firm that will work WITH its customer to achieve the best results. An owner should take this as an indicator that it is past time to look for another ESCO!

The customer need not, in fact should not, "lose control." Every client can, AND SHOULD, first determine the key elements the organization must have in an energy service contract and the latitude within which it is willing to negotiate. This will be discussed at the end of the chapter. Understandably, an energy service company and the financier have to have some assurance that they can protect their investments and that the savings can be reasonably guaranteed.

Contrary to fears engendered by some, this can be achieved without any negative impact on the work environment. In fact, the contract conditions can, and should, stipulate the provisions that will create an enhanced productive environment for workers and occupants.

Attorneys, who are not comfortable with the performance contracting concept, can be a major impediment to achieving an agreement. For them, a piece of the puzzle may be missing. If your organization's attorney does not normally provide counsel in contract law, it is prudent to seek additional or outside counsel. Even though performance contracting has become a more common approach to EE financing, this type of contract may still be without precedent in the attorney's experience. So, even if the lawyer has contract law experience, it will expedite the process if he or she is provided references of attorneys of record who have successfully implemented performance contracts. The state energy office, the organization's consultant, or the ESCO can usually supply such information. It will also facilitate the process if the attorney is brought into the process early by providing him or her a sample energy services agreement and planning agreement to review.

CONTRACTS: LAYING THE GROUNDWORK

Establishing the criteria, preparing the RFP and evaluating the proposals should lay much of the groundwork for the contract. Neither the solicitation document nor the proposal should be considered as all-inclusive, OR BINDING. One US Department of Energy official insisted that no new conditions could be considered that had not been previously introduced in the RFP or the proposal. If ever faced with such bureaucratic "mumbo-jumbo," it may pay to dissuade anyone believing such an "official" by bringing in state or local attorneys with some experience in such matters.

Items not in the solicitation or proposal can be placed on the table for discussion during negotiations. Modifications in what the organization asks for, or the firm proposes to do, are commonplace. Should the proposal or parts of it, by reference, become part of the contract, a statement should be included in the contract indicating the order in which related contract documents prevail.

The solicitation can be used to state that you reserve the right to make the proposal part of the contract. All organizational conditions upon which the proposal was based, however, should remain consistent or the ESCO should not be expected to comply with the stated provisions.

It is strongly recommended that any request for qualifications (RFQs) or proposals (RFPs), ask for a copy of a contract recently executed with a similar organization by the proposing ESCO. The contracts submitted by the ESCOs should be reviewed to get a sense of what the firms really expect. Contracts from the two or three ESCOs under final consideration should also be reviewed carefully by the customer's legal counsel.

There are typically three contracts in a performance contracting "package." They are the Planning Agreement, Energy Services Agreement (ESA) and the financial agreements. The financial agreements are not unique to performance contracting and, in fact, are common to any financing arrangement; so they are not treated here.

Planning Agreement

Brutal experience has taught ESCOs that they cannot afford to give away their energy audits. At one time, the audits were thought to be an effective sales tool. They were certainly effective in showing the owner the existing savings potential, but the ESCO did not necessarily get the work.

As discussed in Chapter 11, an auditor, who can perform the investment grade audits (IGA) which are basic to successful performance contracting, are still in short supply. ESCOs are in the business of selling projects; not audits. An IGA, which does not evolve into a project, denies the ESCO its auditor's time—time which could have been used by the ESCO to open up a real project opportunity.

Owners have also been on a learning curve and recognize an IGA as a premium grade audit that lays the foundation for guaranteed results. Rather than the usual "snapshot" approach, which assumes existing conditions will remain the same, an IGA will give the owner a better understanding of the way certain energy efficiency measures will behave over time in his or her facility. An owner, who can get an IGA for the price of a traditional energy audit, is smart to do so.

To protect the ESCO's interests and to preserve the auditor's time for IGAs that will lead to projects, the Planning Agreement has been introduced into the performance contracting process. The Planning Agreement, also referred to as a project development agreement, is a short contract of three or four pages, which addresses:

a) the objectives which have been agreed to by both parties;

b) the conditions the IGA must satisfy for the owner;

c) a statement that if the objectives and IGA conditions are met and the project is not forthcoming, the owner will pay the ESCO a specified amount for the audit;

d) a statement that if the IGA does not meet the agreed upon objectives and conditions, the owner, of course, pays little or nothing; and

e) the cost of the audit will be rolled into project costs if the project moves forward.

The objectives, which are listed to protect the owner's interest, usually describes the working environment that is necessary and specifies that the recommended measures must not have a deleterious impact on that environment.

The audit conditions in the Planning Agreement generally stipulate the expected range of savings as well as any audit procedures and parameters that are key to the facility manager and owner. A sample Planning Agreement is presented in Appendix C.

Since the IGA is a superior audit and serves as a critical bridge to contract implementation, the IGA cost, as specified in a Planning Agreement, normally carries a premium over the cost of a traditional audit. That premium can be as high as 50 percent above the typical energy audit cost.

The Planning Agreement typically does not require the ESCO to engage in the project. Obviously, the ESCO has pre-qualified the customer and believes a good opportunity exists or it would not tie up its engineer's time, nor incur the costs. Even with the best pre-quals, however, nasty surprises do emerge and ESCOs should not be obligated to proceed with a project that is not economically viable. Under such conditions (if all other provisions are met), the owner usually pays for the audit but at a reduced fee as provided in the Planning Agreement.

Energy Services Agreement

A properly prepared and executed contract assures that a project will move forward with minimum misunderstanding between the ESCO and the customer. If the language is clear and well understood by both parties, and if the terms are fair to both sides, the foundation exists for a cooperative effort that will benefit both the ESCO and the customer. A poor contract invites controversy and bad feelings, often leading to project failure.

Topics generally addressed in an energy service agreement (ESA) are:

- financial terms and conditions;
- equipment/building modifications and services;
- user and ESCO responsibilities;
- risk parameters, such as default/remedies; and
- typical construction contract provisions.

These items may all be covered in one comprehensive document; or, separate schedules pertaining to work in specific buildings, or clusters of buildings, may be added to the contract as work progresses. The one document approach is generally reserved for small projects of limited scope.

Generally, an energy service contract is divided into two parts. The first part of an Energy Services Agreement (ESA) typically states that the two parties agree the firm will supply services to the customer and, in broad terms, outlines the services. And, of course, describes payment for these services.

Typically an ESA is the basic contract and serves as an agreement to agree for multi-phase projects. Attached to the ESA are "schedules" (may also be referred to as attachments, annexes or addenda) which, when agreed to, spell out exactly what is to be done in specific buildings or processes. The schedules become a part of the contract and describe measure-specific conditions, such as how savings are calculated. With a large campus or installation, these schedules may be repeated for each phase of the project. For example, 40 buildings may be involved in the total project but the plan is to do it in phases of five buildings each. A set of schedules will then be developed for each set of five buildings. These schedules are signed by both parties for each phase as they progress.

A tax-exempt organization may enter directly into a installment/purchase agreement with the financier, so it can use its tax-exempt status to obtain the equipment at a lower interest rate. Third party financing is typically used in a guaranteed savings model in North America; the ESCO carrying only the performance risk and not the credit risk. This arrangement with a third party financier is a common approach, which can work with private sector customers as well. Another project financing model is emerging where the lender and/or ESCO create a Single Purpose Entity (SPE), which carries the credit and to some extent keeps the credit off the ESCO's books. In all cases, a parallel agreement with the ESCO can then be entered into to install and maintain the equipment, provide other services and guarantee that savings will cover required payments.

The term, performance contracting, emerged as the accepted label since the energy service company must PERFORM to a certain standard (level of savings) as a condition of payment. These performance considerations are integral to the contract components and are implied throughout most contract provisions.

TYPICAL ENERGY SERVICE AGREEMENT COMPONENTS

1. Recitals (traditional, but not essential)

2. Equipment considerations
   - ownership
   - useful life
   - installation
   - access
   - service and maintenance
   - standards of service
   - malfunctions and emergencies
   - upgrading or altering equipment
   - actions by end-user
   - damage to or destruction of equipment

3. Other rights related to ownership

4. Commencement date and term renewal provisions

5. Compensation and billing procedures

6. Baseyear conditions/calculations, baseline adjustment provisions, and a re-open clause

7. Measurement and verification procedures
   [In multi-phase projects, items 6 and 7 are addressed in the schedules and general terms of reference are used in the master contract.]

8. Late payment provisions

9. Energy usage records and data

10. Purchase options; buyout conditions

11. Insurance

12. Taxes, licensing costs

13. Provisions for early termination
    — by organization
    — by firm
    — events and remedies
    — non-appropriations language (for government entities)

14. Material changes

15. Conditions beyond the control of either party (force majeure)

16. Default
    — by organization
    — by ESCO

17. Events and remedies
    — by organization
    — by ESCO

18. Indemnification
    — for both parties

19. Arbitration

20. Representations and warranties

21. Compliance with laws and standards of practice

22. Assignment

23. Additional contract management terms
    - applicable law
    - complete agreement
    - no waiver
    - severability
    - further documents, schedules

24. Schedules (by designated group of buildings, or project phases)
    - description of premises; inventory of equipment
    - energy conservation and/or efficiency measures to be performed
    - baseyear conditions and calculations, variables and baseline adjustment provisions
    - savings calculations; formulas
    - measurement and savings verification procedures
    - projected compensations and guarantees
    - comfort standards
    - contractor O&M responsibilities
    - O&M responsibilities of the owner
    - termination, default value, buyout option
    - treatment of existing service agreements
    - calculation of other savings; e.g., existing service/maintenance contracts
    - contractor training provisions
    - construction schedule
    - approved vendors/equipment

A contract offered by an ESCO is designed to ensure that its interests are protected. It is not necessarily designed to protect the interests of the owner. As in all contract negotiations, it is up to the customer to make sure its interests are protected. Prior to negotiating a contract, both parties need to consider the implications of the various key components and the latitudes within which an item can be negotiated. In other words, decide what is not negotiable, what conditions can serve as "trading stock" and in what priority.

The best contracts are tailor-made for a specific client and a particular project. No one "model contract" will suffice. The key contract considerations offered below provide a foundation for developing the contract.

KEY CONTRACT CONSIDERATIONS

Any good performance contract will provide clear understanding of each key contract element. The following discussion offers a context in which these key matters can be considered with an attorney.

Equipment Ownership

The financing scheme used and the point at which the organization takes ownership can affect the organization's net financial benefit and may affect depreciation benefits.

In an energy saving project, the useful life of the proposed equipment is a key factor in post-contract benefits.

ESCOs and/or their financiers usually insist on a first security interest in the installed equipment or collateral of equivalent value.

In the case of buy-out provisions, termination and default values, procedures for establishing capitalized equipment worth may be set forth in the original contract. Terms, such as fair market value, need to be carefully defined. The buy-out provisions will typically be greater than the value of the equipment, as the ESCO's fees for services, risks and potential savings benefits need to factored in.

Malfunctions

Provisions for immediate, and back-up, service in the case of malfunctions need to be spelled out. This is especially important if the contractor is not a local firm. Local distributors for the selected equipment frequently serve this function with further back-up provided by the ESCO. Maximum downtime needs to be considered. The allowable emergency response time will vary with the equipment installed and how essential it is to the operation.

ESCOs need to establish an understanding with the distributor or designated emergency service provider as to the timing and the extent to which emergency services will be provided before they are committed in the performance contract.

Firm Actions, Damage

Contracts proffered by ESCOs will discuss actions the customer might take that could have a negative effect on savings. The management needs to determine if these conditions are reasonable and determine to what extent the organization should have the same protections.

Consideration should also be given to the impact the ESCO's redress may have on the organization. If termination provisions provide for equipment removal, the conditions of the facility after that removal and the consequences for facility operations need to be stated and understood.

Equipment Selection and Installation

The customer should reserve approval rights on selected equipment provided approval is not "unreasonably" withheld.

ESCOs must retain some rights if they are to guarantee the savings. Under some bid procedures, the energy service company may take on the role of a general contractor: writing specs, monitoring bid procedures, and overseeing installation. These may be services a given organization needs, but they also serve to protect the ESCO's position on guarantees. In any case, with guarantees involved, an ESCO must retain sufficient control of the specs as well as equipment and installer selection to assure guarantees can be met.

To effectively meet the concerns of both parties, equipment specifications are generally developed cooperatively. Often, one party then narrows the selections to two or three bidders and the other party makes the final selection.

Contractual conditions used in any construction project; i.e., liability, OSHA compliance, clean up, performance bonds, etc., should apply.

Provisions for Early Termination

From the customer's point of view, contract language regarding termination should clearly set the parameters for equipment removal. These provisions should include length of time the removal will require and provide sufficient details as to restoration of the facility after the removal.

ESCOs incur major exposure early in the contract, for they incur the major expenses at this time and must depend on eventual savings to cover these costs. Buy-out provisions must provide for ESCO recovery of costs incurred and a proportionate profit. Buy-outs may not be offered as an option until a specified period, as long as two years, has elapsed.

For further protection, ESCOs, or their financiers, frequently specify that a tax-exempt organization use non-appropriations language in which the owners agrees not to replace the equipment with equivalent equipment within a specified time frame.

Conditions Beyond the Control of the Parties

Usual contract language absolves the ESCO of certain contract responsibilities under force majeure, or acts of God. These conditions should be examined, and the merits of similar provisions for the organization should also be weighed. Increasingly, the language is written to absolve both parties equally.

Default Language

Language frequently limits the conditions of default and remedies for the ESCO, but may leave it wide open for the customer. When the financial burden is carried by the ESCO, this is not necessarily inappropriate. Similar language for the organization should be considered, especially if the organization holds the debt service contract on the equipment.

The owner's attorney should carefully consider the default and remedies it should impose on the ESCO and accept for the client.

Indemnification

Both the ESCO and the customer should be indemnified. Some ESCOs attempt to secure indemnification from indirect and contingency damages. These are frequently too broad and should be analyzed carefully by the organization's attorney.

Assignment

The customer should insist on prior approval for any assignment, including any changes of service responsibility, or key personnel. Prior approval of subcontractors is also desirable.

Applicable Law

The ESCO typically presents a printed contract as the basis for agreement. The ESCO is apt to specify the applicable laws of the state in which it is incorporated. Should court action be necessary, the ESCO has a cost advantage and possibly a legal advantage. This places an additional burden on the customer if the organization is located a state different from the one specified in the contract. Since applicable law provisions may just as easily specify the customer's state, this provision may become "trading stock" in the negotiation process.

Savings Calculations Formulas

This procedure is frequently made far more complex than it needs to be. The reduction in units of fuel and electricity multiplied by the current cost of energy by unit is the standard procedure for calculating cost of saved energy. Attribution of electrical demand charge savings also needs to be evaluated, negotiated and specified in the contract language.

Weather or occupancy changes, added computers, etc., can effect savings. Extensive contract language, however, which tries to anticipate every contingency, which might happen, only benefits the legal profession. The clearest way to address this problem is to have a broad-based baseyear and annual adjustment provisions related to anticipated variables, which both parties approve. Then the parties should agree on a re-open provisions, which in effect says, "If the annual baseline after adjustment differs from the baseyear by more than ± "X" percent, the baseyear provision can be re-opened to negotiate a new baseyear reflective of current conditions." It is important that it be stipulated that only this one section can be re-opened unless opening a can of worms has appeal.

Calculation of Baseyear and Adjusted Baseline

Provisions for calculating a baseyear should be clearly presented. Historically, baseyears were often established simply by averaging multiple years of consumption. That is not enough! In all cases, conditions that have driven the consumption should noted and any potential changes, which will have a major effect on consumption should be clearly identified. Baseyear consideration should include; (1) mild or severe weather in recent years, (2) recent changes in the structure, building function, occupancy, etc.; (3) recent O&M work which significantly affect consumption; and (4) any recent renovation which could impact energy consumption. Reopen language should provide for some adjustment beyond the agreed upon variations; so neither party pays for unexpected contingencies, such as the closure of a wing of the facility or added computer labs.

The share of the savings will vary with the length of payback, the services delivered, the financing scheme selected, the risks assigned to the

ESCO, the length of contract, and the like. The interrelationship of these factors needs to be considered in negotiating each party's share of the savings.

Measurement and Verification

Procedures for measurement and verification (M&V) of savings will vary with the energy efficiency measures installed, the size of the project and a number of other factors. In the ESA, the contract language typically states that the M&V procedures will be decided jointly, following the IGA and according to the International Performance Measurement and Verification Protocol (IPMVP). Copies of the most recent protocol can be download at no cost at the IPMVP web site: *ipmvp.org*. Further discussion of key M&V considerations are presented by Dr. Roosa in Chapter 5.

Energy Prices

Price volatility needs to be given careful thought. How the "burden" of falling prices or the "benefit" of rising prices is to be shared should be clearly addressed in the contract. Since no one should ever be foolish enough to predict energy prices two, five or twenty years out, ESCOs should insist on a floor price. In other words, an ESCO should not only be able to guarantee the amount of energy saved, but should also guarantee the value of the energy saved will be sufficient to meet the customer's debt service obligations *provided the price of energy does not go below a set floor price*. If the ESCO insists on a price floor, however, the customer should enjoy the benefits associated with increased energy prices.

Comfort Standards

The greatest fear employees associate with energy efficiency, and more particularly performance contracting, is the loss of control of the work environment, particularly comfort factors. The frequently voiced supposition that an energy service company will control the building operation is simply not warranted—unless management abrogates its responsibility and gives the control away.

The management can, and should, establish contractually acceptable comfort parameters for temperature, lighting levels and air exchange as well as the degree of building level control needed (and override required) to assure a quality environment. The owner's latitude of control can reduce savings. In such instances, the ESCO's risks becomes greater. The more control the customer keeps, the greater the ESCO risks and the lower the customer's share of the savings will be. Indirectly customers always pay for actions which increase ESCO risks. (See chapter 10 for a more detailed explanation.)

Projected Compensation and Guarantees

The most attractive part of performance contracting is the idea that there is an entity out there that will make sure the organization has new capital equipment THAT WORKS, and can assure that the savings will occur and all this will happen without any initial capital cost to the organization. The manner in which the energy savings are guaranteed to cover debt service payments, and the risks associated with meeting that guarantee, are key components of a performance contract and deserves careful consideration.

Since the quality of maintenance on energy consuming equipment affects savings, most ESCOs require specified maintenance provisions and may ask for related maintenance contracts. They may, however, not guarantee that energy savings will cover the required maintenance fee. If an organization regularly contracts for maintenance and the ESCO's fee is not greater than the existing fee, this may not pose a problem.

A major reason for a contract is to identify and assign risks and provide appropriate recompense. The "guarantees" are the bottom line in making sure a contract works in the organization's favor. However, the more risks accepted by the ESCO, the lower the savings benefits will be to the customer. As discussed in Section II, *money always follows risk.*

As with any contract, your attorney should review the ESA before signing. Through all the negotiations, frustrations and delays, it's well to remember that a good contract is essential to a successful project.

SECTION-BY-SECTION

Model contracts are frequently requested, but that can prove dangerous if the contract is not carefully modified to meet the unique conditions of state laws, local ordinances and customer conditions. The following section-by-section analysis is provided with the understanding that local attorneys can then develop a contract that specifically meets an organization's concerns.

INTRODUCTION (RECITALS). The opening section identifies the contractual partners, states that both organizations are in business and then briefly states why they wish to enter into a contract.

SECTION 1. **Energy efficiency program**. This section describes in broad terms what will be done on the customer's property and usually lists the schedules that will be attached which detail the actual work, savings formulas and other matters that are measure specific.

SECTION 2. **Customer's energy usage records and data**. This language states that the customer will make available in a timely manner the necessary information about energy use and other data needed to calculate potential savings and measure actual savings.

SECTION 3. **Commencement Date and terms**. The calendar dates when the contract will take effect and ends are in this section.

SECTION 4. **Payments to ESCO and customer**. Language here provides for the customer to pay for the services that will be spelled out in detail in the attached schedules. The schedules will contain the formulas by which savings are calculated and the way savings may be divided between the ESCO and the customer.

SECTION 5. **Coordination**. This section simply states that the ESCO will not cause unwarranted interference with the business of the customer during the installation of the project, and that the customer will cooperate during the installation phase.

SECTION 6. **Ownership**. Section 6 establishes the ownership of installed equipment, spells out the ownership rights and the conditions and timing of title transfer. It also includes any title provisions should the contract be terminated.

SECTION 7. **Upgrading, altering, removal or damage** of installed equipment or systems is covered in this section. Because the ESCO depends upon the correct operation of the installed equipment to produce savings, this section limits what the customer can do should such actions change or modify that equipment. It usually also addresses what happens when the system is damaged.

In addition, this section states that the ESCO may, with the prior approval of the owner, upgrade or improve installed systems if savings will be enhanced.

SECTIONS 9 & 10. **Material change**. These sections address what happens if the organization makes substantial changes to its facilities (or closes a facility) which alter the energy situation during the life of the contract. A second section provision generally treats notification procedures in the event of material changes.

SECTION 11. **Insurance**. Requirements for insurance are similar protection to that required on any construction project.

SECTIONS 12 & 13. **Conditions beyond control of the parties**. These force majeure sections address matters beyond the control of the parties, such as floods, earthquakes and other acts of God, which may disrupt the project or significantly alter savings potential.

SECTIONS 14 & 15. **Defaults and Remedies**. These sections discuss what happens if either party fails to live up to the terms of the agreement. The first section specifies what constitutes a default by either party. The second section identifies the remedies available to each party.

SECTION 16 **Termination**. This section establishes the means by which the ESCO or the customer may terminate a contract, and may set forth the terms under which a customer may "buy out" a contract before the ending date.

SECTION 17 & 18. **Indemnification and Arbitration**. The first of these is a standard "hold harmless" clause in which each party will be equally protected. The second section suggests how disputes between the parties should be handled through arbitration.

Several "housekeeping" and contract management sections generally follow assuring that the parties have the authority to sign contracts, that what they have said in the contract is true and that the contract complies with local laws and standard practices.

SCHEDULES. This section, often located just before the signatures, states that the schedules to be attached are a part of the contract with a statement that, in case of conflicting provisions, the text in the body of the ESA will prevail. (Generally the schedules detailing the project are prepared, and negotiated, after the general agreement is signed.)

The Schedules

The schedules attached to the general ESA serve to make the contract specific to the project, or a phase of the project. They establish the details of the work to be done and the conditions under which the work will be accomplished. They include the manner in which savings will be calculated, measured and verified, and services that will be provided for specific measures.

The content and number of schedules may vary. Usually the following types of questions must be addressed, each of which may be the subject of negotiation:

A. **Equipment**. What equipment exists? What equipment will be installed? What is the projected cost of the equipment? Who will install it? Who will maintain the equipment?

Building modifications are usually addressed in this schedule as well.

B. **Warranties**. How and when will the manufacturer's warranties be conveyed to the owner? How will the maintenance tasks be monitored? Some contracts couple guarantees with warranties.

C. **Savings formula**. What are the assumptions and formulae that are the basis for the energy savings calculations? Allocation of demand charge savings and positive cash flow are treated here. Changes prompted by utility restructuring; i.e., real time pricing, also need to be addressed.

D. **Measurement and Verification**. Are the savings to be stipulated, and/or measured? How will the savings be measured and verified? Actual procedures, measurement devices and assigned responsibilities are set forth varying with the measures to be installed.

E. **Guarantees**. What are the guaranteed savings by year? What are the payments to the ESCO from the savings by year? What are the guarantees by the ESCO to the customer? What are the guarantees, if any, by the customer to the ESCO? What are the procedures to adjust the baseline for reconciliation? When will the reconciliation occur? [In organizations which operate on a fiscal year it saves a lot of budgetary grief if the reconciliation is done just before the end of the fiscal year.]

F. **Baseyear**. What were the consumption and operating conditions of the customer's facilities, process and equipment prior to project retrofits? What operating conditions and/or assumptions are used in the calculations? How will the baseyear be adjusted to accommodate predetermined variables?

G. **Price variation**. When costs vary due to inflation or other factors, especially energy prices, what happens? Is there a floor price? How are price increases shared?

H. **Performance standards**. What customer operating performance standards must be met by the improvements installed by the ESCO; e.g., lighting conditions, acceptable temperature ranges, steam flow, etc.? What are the equipment installation schedules? And what standards of service must be met?

I. **Ownership**. Who will own the equipment during the life of the project? If the customer wants to purchase the equipment during the project earlier than planned, what are the terms and conditions of the purchase?

NEGOTIATIONS

Effective negotiations lead to effective contracts. And more importantly, effective contracts make for good projects. Ideally, when the negotiations are over, all parties should walk away from the table feeling they have laid the foundation for a strong partnership—one which will serve everyone well for many years.

Before negotiations start, each party should take stock of its own operation and what strengths it brings to the table. Careful consideration should be given to what conditions are negotiable and how much latitude can be allowed on the negotiable items.

It is also important to learn what one can about the other party. Not just what is known that brought you to this point, but what is the other party's negotiating history and behavior at the table. A couple of phone calls in advance can often prevent some surprises.

Both parties will profit from a little self-interrogation. What do "they" offer that you must have? What does your organization offer that is particularly attractive to them? What is the best way to position your strengths in the discussion?

Know the process. Ignorance can weigh heavily against you. Negotiations have some uniformity regardless of the topic. Recall previous negotiations, even union negotiations. Consider what your strategies will be, and anticipate what kind of strategies can be expected from the other party.

NEGOTIATING STRATEGIES

Careful thought regarding some very basic negotiating strategies can make you and your organization feel more comfortable, as you head for the table.

The suggestions offered below have been roughly drawn from *Roger Dawson's Secrets of Power Negotiating.* For those with limited negotiation experience, this book is strongly recommended for use as a guide.

1. The customer should carefully review the sample contract submitted with the proposal before the ESCO is initially selected.

2. The customer's attorney should meet with those negotiating the contract prior to negotiations and go over the draft contract submitted by the ESCO. Then, the group can:
   a) set-aside the parts which are acceptable;
   b) note those parts that need slight modifications;
   c) note those parts that might be key to the ESCO, but not necessarily to the customer;
   d) identify the parts which are unacceptable and what needs to be changed—and how—to make it acceptable; and
   e) decide just how much latitude there is on each item and what other parts have some "give" to be sure the key parts are developed to your liking.

3. Never accept their first offer. Even "printed" contracts can, and are, revised.

4. Ask for more than you expect to get. The other party assumes you will. Starting where you wish to end up, too often leads to getting less than you wanted—or should have.

5. Avoid confrontational negotiation. The other party will be your partner for many years; so start as you mean to go.

6. Display some traits of the reluctant buyer/seller as part of your strategy. Eagerness has its place, but seldom at the negotiating table.

7. Reserve the right to defer to a higher authority; i.e., the boss or the attorney. Generally, attorneys complicate things and too often want to get into legalese. Whenever possible, it pays to keep them out of the negotiation's room. (Remember, attorneys will not be living daily with the project.) Attorneys, however, can be very useful out of the room as the "higher authority."

8. Remove *their* resorting to a higher authority by appealing to their egos or pressing for them to commit to making a recommendation of a certain position to that authority.

9. Be on the look out for their "problem," which you can help solve. Recognize it as a "hot potato" and test its validity.

10. Never, ever, offer to split the differences, but you might encourage them to do so.

11. A critical point, which has been noted by Mr. Dawson, is particularly important for performance contracting: perceived values during negotiations go up for materials and DOWN FOR SERVICES. Protracted negotiations can, therefore, diminish the perceived value of services. Considering that performance contracting has a strong service focus, the negotiation process can work for or against a party depending on whether you are buying or selling those services.

12. There are two basic rules on making concessions:
    a) always get something in return; and
    b) start big and taper off.

    If your concessions get bigger as you go, the rewards for the other party to continue negotiations are obvious.

13. If you truly reach an impasse, consider setting it aside to deal with later.

14. Should you reach deadlock on a key issue, consider intervention or mediation.

15. Position a point for easy acceptance by leaving something on the table.

16. Watch out for the "Oh, by the way" when it seems the negotiations are over and everyone is smiling and shaking hands. This last little "nibble" could be bigger than it seems.

17. Never lose sight of the fact that a good contract is one where both parties feel they have a fair and workable deal.

If I Were on "Their" Side of The Table

To balance the scales, the customer should picture himself/herself on the ESCO's side of the table. When guarantees are part of the picture and performance is tied to the guarantees, there are some items that are virtually non-negotiable for the ESCO. A county once put out an RFP that glibly stated that the ESCO would carry the financing, make the guarantees and the county would select the equipment. Surprisingly, they got several responses; not so surprisingly most were from very new ESCOs. Happily several conditions were changed before any contracts were signed.

In order to make a guarantee on the savings from the project, an ESCO will expect to:

a) write the specs in cooperation with owner and participate in the final selection; or

b) select the equipment with owner final approval;
c) select the subcontractors who will install the equipment with owner's tacit approval; and
d) decree the level of maintenance and tasks to be performed by the owner with some key maintenance provisions reserved to the ESCO.

If the customer feels a strong need to have control of any of these items, they can expect the ESCO to hold back a large financial cushion to cover the risks. This step, in turn, leaves the customer with a smaller project, less savings and fewer overall benefits. Occasionally the alternative is to remove the guarantee provision related to this item, or entirely from the contract.

If the guarantee that savings will cover the debt service obligation is removed, some performance conditions can still be maintained by developing a shared savings model for the excess savings. In this scenario, there is no assurance that the savings will cover the debt service obligation, but the owner is somewhat assured of the ESCO's continued interest in the project's performance by splitting any savings over and above the debt service payment.

Reference

Dawson, Roger, *Secrets of Power Negotiating.* 1995. Career Press, Hawthorne, New Jersey.

Chapter 8

Where the Savings are: Effective Project Management

A surprising number of ESCOs do not pay sufficient attention to the management of a project once installed. Ironically, that's the profit center for the ESCO and the customer. To gain maximum benefit, the customer, whose vision is too often clouded with new equipment hopes, should insist on it.

From the ESCO point of view, the Project Manager is there to make sure the customer is happy, the project is operating properly, resale opportunities are identified and pursued, and *expected savings are realized.* For the customer, the Project Manager is the go-to-guy, who can resolve any project-related problems, maybe solve new problems, make sure all project-related concerns are heard and addressed by the ESCO, and that *expected savings are realized.*

If the connection is not made, the savings will not be realized. Then, the ESCO and the owner are in for a shock.

The Project Manager is THE liaison—THE conduit. At the other end of that conduit is the Energy Manager. If either end is not open to dialogue, an information bottleneck results and the project suffers. They must work together effectively to assure project success. (See Chapter 9 for a discussion of the critical role of communication in project management.)

A small matter of semantics probably needs to be cleared up at this point. In construction work, the terms *Project Manager* and *Construction Manager* are frequently used interchangeably. In performance contracting a distinction is made. The Project Manager overseas the project for the life of the contract while the Construction Manager is responsible for the installation of the equipment, process modifications and building envelop changes. The Construction Manager's responsibilities are essentially done at the point where the owner signs the Certificate of Acceptance.

The Project Manager needs to keep a focus on project energy savings. For the ESCO, this is what the project is about—where the ESCO's profit is. Both the project and Project Manager's success rests on delivering savings day after day, month after month and year after year through the life of the project.

In this chapter, we will, therefore, first consider project performance from the ESCO's perspective and the role of the Project Manager. Then, we will view the project from the owner's and Energy Manager's perspective. The critical role the Project Manager and the Energy Manager play will become very evident.

The long-term partnership, which is basic to performance contracting, can be a tremendous advantage to both parties; but *only if it is treated by each party as a long-term partnership*. Effective cooperation can mean more savings, more immediate problem resolution, and simply more effective projects. The focal points of that partnership and its success is the responsibility of the customer's Energy Manager and the ESCO's Project Manager.

Those energy service companies, which have become, or remain, successful in the 21st Century, recognize a strong partnership with the customer as a critical component of any effective savings-based agreement. The level of commitment that is exhibited by the customer's management, therefore, becomes a decisive factor as to whether or not an ESCO will invest in a certain industrial or commercial facility, university, hospital, or government building.

So, how do we achieve effective project management? The critical components are sound management and effective communication.

ESCO MANAGEMENT STRATEGIES

A "hit and run" approach seldom works in any business arrangement. It definitely won't work when an agreement requires performance over the life of the contract. ESCOs that thought all they had to do, once a sale was complete, was install the equipment and issue a monthly bill are weaker by the day or no longer with us. The industry is full of horror stories that can be traced to poor ESCO leadership, inadequate communications and poor project follow through between the customer and the ESCO.

A working partnership will require more than lip service. The effective ESCO looks at the organization's needs and serves them. Service that is directly related to energy savings should not be farmed out. For example, one ESCO in the 1980s contracted out all of its monitoring, billing and maintenance on existing contracts for a set fee. The ESCO lost contact with its customers. The subcontractor had no incentive to provide the quality of service that meant greater savings. Everybody lost.

Everyone involved, especially subcontractors and Project Managers, should have a vested interest in project success. This means the "pay for performance" concept should be pervasive at all levels.

The best performance contracts today are much more than financing. They offer specialized energy expertise, improved capital equipment, training, monitoring—whatever the customer wants that can be reasonably built into the package. In fact, the effective ESCOs are increasingly laying out a "smorgasbord" of opportunities from which customers can pick and choose services that most precisely meet their needs, and which the ESCO can demonstrate provide added value.

ESCOs, which expect to endure through the years, must envision progress up the value chain to the point where the ESCO can offer the customer business solutions. Those needs may not include financing. They may, however, include guarantees that enhance the customer's ability to do its own financing.

The performance contracting process also requires an attorney, who knows something about contract law, financing, and energy efficient technology. Performance contracting is still a relatively new concept and not well understood by all attorneys—and certainly not in all countries. Too often, rather than acknowledge their limitations, lawyers drag out the decision-making process or advise against the whole concept. ESCO sales personnel and management must be all things to all people, or draw on independent third party financial, legal and technical contacts to educate their customer's counterparts.

The successful ESCO anticipates what the Energy Manager needs in order to maintain an informed staff and management. Then, the ESCO makes certain that the Energy Manager has the information when it's needed and in a useful form. Such an effort not only maintains commitment and enhances savings; it paves the way for contract enhancement and contract renewal.

THE PROJECT MANAGER

To maximize project success, it is vital for each ESCO project to have a Project Manager. Whether the Project Manager is located on-site or makes regular site visits will likely be determined by the project scope, size, range of services, and other factors of the performance contract. But for the comfort of the customer and the coherency of the project, a single point of contact, a voice for the ESCO with the client, is essential.

There are several organizational and operational conditions, which must be present in order to make the most of the opportunities of performance contracting: Some should serve as criteria in customer selection. Once the customer is selected, it is the Project Manager's job to be sure the following objectives continue to be met effectively:

1. Endorsement by management in the form of an organizational statement or policy… a demonstrable level of commitment;

2. Coordination of project tasks and designated responsibilities through central positions: the owner's Energy Manager and the ESCO's Project Manager;

3. Involvement and support of the occupants, especially operations and maintenance personnel;

4. Identification of internal and external resources needed to do the job; ways to smoothly access this support;

5. On-going project assessment; both project evaluation and M&V;

6. Appropriate attention to details and schedules to meet project objectives; and

7. A carefully designed and implemented communication strategy.

Performance contracting requires effective leadership in each of these areas by both the ESCO AND the customer. Exercising that leadership in the customer's organization is usually more complex than in the ESCO, primarily because it often involves many people at many organizational levels. Frequently, these people do not have energy concerns as a major responsibility. It is not unusual to find a disconnect between the Energy Manager and top management. All too often, they even seem to be working against each other. The Project Manager must recognize and support this range of Energy Management needs.

Selection of a Project Manager does not mean simply hiring or assigning the best technical person for the job. If this is to be a partnership between the customer and the ESCO, the Project Manager will be key to the partnership. His or her selection should be based on ability, technical management expertise, and communication skills—with some common sense thrown in about how the ESCO relates to the customer. Conditions in the customer's organization will also influence the selection. A good Project Manager will walk a fine line between protecting the ESCO's investment and being an advocate for the needs of the client.

If given a choice in Project Manager selection between a highly skilled technician and a good leader, take the leader. The needed level of technical expertise can be taught, but unfortunately strong leadership skills are very difficult for some people to acquire.

The fact is, however, that even if a Project Manager has the best skills, but not the ability to work within a particular environment—from urban to rural, from one part of the country to another—those skills may be useless. A psychological/social fit is also a part of the equation. This becomes particularly important when an ESCO is working nationally or globally. Insensitivity to social and cultural norms can ruin a project, kill project profitability, and destroy opportunities for future projects. Even with the increasing homogeneity in the US, a New York Project Manager, who really knew his job, almost ruined a project in southern Georgia because culturally he was a misfit.

In this role, the Project Manager will likely be most effective if he or she understands that winning over people within the client's organization, who might be skeptical of new processes, will make everyone's life easier and the project more successful. Informing the operations and maintenance staff about their new roles or new equipment, without achieving their "buy-in" can be deadly. Many a project has been sabotaged because it wasn't considered important to include maintenance staff in project development

Finally, the Project Manager for the ESCO has to have the authority to adjust the project as unexpected situations arise. Assigning a Project Manager, who must always "check back," will diminish his or her stature with the customer and could reduce project potential.

An effective Project Manager will:

- be involved as part of the initial IGA risk assessment;
- during initial evaluation, help determine O&M attitudes as well as training and manpower needs;
- merge the technical people in-house and outside support, into a smooth working team;
- help manage effective energy efficiency communications within the organization;
- serve as a liaison between the customer and the ESCO; representing the concerns of each entity to the other;

- identify and expeditiously resolve any problems in a cooperative and collegial manner; and

- become an adjunct to the effective and efficient operation of the customer's organization—even finding business solutions before the management knows it has a problem (which may not even be related to energy efficiency).

An effective Project Manager often finds upgrade potential or business solution opportunities over the life of the contract that exceed the ESCO's initial investment.

There are many subtleties in effective project management. A key one is assessing the Energy Manager's stature within his/her organization, and determining what the Energy Manager can accomplish. Project Managers should not ask more of an Energy Manager (EM) than he/she can deliver.

During the life of the project, many opportunities will emerge, which might enhance the Energy Managers position in the company. Many of those opportunities can be initiated or enhanced by the Project Manager (PM). When things go well, it is key that the Energy Manager gets credit.

It's amazing what one can accomplish if you don't care who takes the credit.

The most effective PM/EM relationship will emerge when both the PM and EM see their counterparts as equals in stature and influence. Unevenness can result in the dominance of one or the other; resulting in a loss of balance in the partnership, which in turn can adversely affect the project.

Only in a dire emergency should the PM by-pass the EM and go to a higher level of authority in the customer's organization. This by-pass can diminish the EM's stature (real or perceived) in the company and jeopardize the PM/EM working relationship. If the PM must go to a higher level, every avenue should be explored to include the EM in this effort. The only exception to this rule is when the EM needs to be removed from his/her position. Careful deliberation by and within the ESCO team should precede any such action.

Both the PM and EM should always remember that each has the power to make the other party look good. Their individual success and project success is dependent on a strong effective working relationship.

MANAGEMENT STRATEGIES FOR THE OWNER

Performance contracting should be a way for an organization to address its energy needs, its equipment needs and its energy-related maintenance concerns. The organization, however, should not assign more to, nor expect more of, its ESCO partner than the ESCO has historically offered its previous customers. Grand and glorious promises, which the ESCO does not have the resources to deliver, pave the way for growing disenchantment that can ultimately lead to court.

A tremendous advantage that accrues to an organization that elects to use performance contracting is the expertise and experience it gains from the ESCO, backed by its guaranteed performance. Unfortunately, this advantage can be so tantalizing that some organizations assume the ESCO can do it all; and, in effect, they abrogate their own responsibilities to manage the process.

The ESCO can bring its expertise to the partnership, but occupant behavior can make all the difference in whether that expertise bears fruit and in the level of savings achieved. Those who abrogate their portion of project responsibility suffer major consequences. Uncommitted administrators, uncaring staff and indifferent operations and maintenance personnel can subvert or undermine the best an ESCO has to offer.

ESCOs seldom, if ever, guarantee the maximum possible savings. A quality investment grade energy audit* will provide a sound estimate of the saving potential under ideal conditions. Understandably, the ESCOs have a vulnerability that causes them to be cautious with their guarantees. The guaranteed savings level is typically up to 80 percent of what an ESCO can reasonably expect a project to realize. As the ESCO assesses the risks associated with a given customer/project, the guarantee is adjusted accordingly. As the risks go up; the guarantee percentage goes down. As the guarantee goes down, EVEN WITH THE SAME LEVEL OF INVESTMENT, the size of the project, the amount of equipment installed and the potential savings all drop off.

The customer's risk profile, therefore, makes a huge difference in how much it will realize from its investment.

It clearly follows that an organization that does not work hand-in-glove with its ESCO can impede the ESCO from reaching even the 80 per-

---

*See *Investment Grade Energy Audits: Making Smart Energy Choices* by Shirley J. Hansen and James W. Brown. Published by the Fairmont Press, 2003.

cent mark. But with a strong supportive partnership, both parties may enjoy a positive cash flow of up to 100% of the predicted potential; sometimes even more.

Putting the "manage" in energy management requires both administrative commitment and leadership. The real difference between an effective energy program—and one that is not—is the attention paid to the people factor. It is up to each member of the management team to mobilize this support. Energy leadership qualities, which draw from the array of management skills already in hand and redefine them within the context of the organization's energy concerns and opportunities, rest primarily with the Energy Manager.*

For most managers, limiting the workday to eight hours sounds like a remote dream. The mere thought of adding to the myriad of administrative responsibilities is enough to make managers shudder. Efficient energy usage, however, is not added to the day. It permeates all activities. Fortunately, once its components have been set in motion, a comprehensive energy management program under ESCO monitoring usually requires only limited attention and reassessment. Visible and continuing attention, however, even if not all consuming, is essential.

Even before management of the project begins, and before the contract is signed, the owner should be aware that an experienced ESCO not only surveys the energy saving potential of a building, but has also assesses the management and staff's potential to support the project. The most attractive potential customers have a written energy policy in place, a designated person responsible for energy efficiency, and a demonstrable commitment to both.

SETTING POLICY

Management can be transitory. Adopting an organizational energy policy cements energy positions, gives enduring guidance and reassures the ESCO that support is stable. Frequently, the support for performance contracting is made a part of the organization's energy policy, integral to a broad statement of administrative commitment to effective energy management.

---

*For suggestions on improving the management function, see the paper "Putting Management into Energy Management" in the WEEC proceedings, 2005; and the *Manual for Intelligent Energy Services*, published by the Fairmont Press, 2002; both by Shirley J. Hansen

Organizations with adopted energy policies are typically viewed by ESCOs as less risky. This perceived reduced risk leads to a stronger guarantee and ultimately more revenue for the customer for the same level of investment.

Typical policy statements include:

- A statement of the mission of the organization and how energy efficiency relates to that mission; i.e., a commitment to make wise use of limited resources, a commitment to implementing all practical ways to increase the bottom line, a commitment to creating a safe and productive work environment for employees;
- A statement of concern regarding the broad energy situation and more particularly economic and supply implications—including energy security concerns—for the organization (and the community);
- A statement recognizing the advisability and cost-effectiveness of developing energy management procedures;
- A statement of commitment to implementation considerations such as;
  - — authorizing the position of Energy Manager,
  - — delegating authority to the Energy Manager within specified parameters,
  - — establishing a budget to support the position and process;
  - — identify funding procedures;
  - — requesting that an energy management plan be developed for board and/or administrative approval; the plan to include goals (potential reductions), energy costs (history patterns/projections), potential savings, suggested procedures, and recommendations; budget, and funding procedures [an investigation and potential implementation of performance contracting may be explicitly mentioned]; and
  - — the critical need for regular evaluation and revisions as appropriate; and

- Reporting requirements to assure the policy is implemented and adhered to, which should incorporate evaluative data and further recommendations.

A policy should be just that, a brief statement of policy. It should not include day-to-day considerations, or the mechanics of implementation. Specific implementation plans, such as temperature settings, belong in an energy management plan.

THE ENERGY MANAGER

If a position has not been designated as the Energy Manager, even for a portion of his or her time, then it is doubtful anyone is managing energy use. A feel-good statement as to how "everyone is responsible" may look good in print, but in practice it means no one is responsible.

For successful performance contract management, one person on the organization's staff should be designated as responsible for energy matters and given the time, the authority, and the budget to do the job.

For maximum effectiveness, the Energy Manager should be in a position on the "org. chart" to effect change. If the Energy Manager is buried in the facilities department, the opportunity to meet critical energy responsibilities are very limited. The focus should be on "manage" and the Energy Manager should be part of the management team. Where the Energy Manager is positioned in an organization also provides critical information to the ESCO on the level of emphasis the organization places on energy use management. *In short, for an organization to have an energy program that makes a difference, the Energy Manager's position must be perceived to be at a level to get the job done*

It should also be noted that the Energy Manager needs to earn his/her right to sit at the management table. If the Energy Manager talks, equipment, technology and Btu all the time, that person will not be regarded as *management*. Energy is the life blood of an organization. It is indispensable to its operation. The person who manages it should be part of management.

Why does an organization retaining an ESCO need an Energy Manager? The role of the Energy Manager may depend on how comprehensive the contract is and how the Energy Manager's role is defined. But, at the very least, part of someone's efforts needs to be devoted to working with the ESCO to make sure the organization and the ESCO are each holding up their end of the partnership.

A guiding rule of thumb for an organization is: if it can put an Energy Manager in place (or dedicate a portion of his/her time) for 10 percent of the annual utility bill, the organization can't afford not to. An effective EM will always find ways to reduce the energy bills by at least that amount.

Specific job descriptions for the Energy Manager position will be situation specific. But, key ingredients of an Energy Manager's job should include:

- setting up and/or implementing an energy management plan;
- developing an energy security plan—either as a separate document or part of the energy management plan;
- establishing, maintaining, and sharing energy records by consumption and cost;
- identifying assistance available from other sources; e.g., utilities, federal/state grants, and exploring ways to leverage such funds or assistance;
- participating in management planning which has energy supply and/or use implications;
- assessing future energy needs; overseeing energy audits;
- convening and staffing an energy committee, which represents a cross section of the organization;
- making energy recommendations, *in conjunction with an energy committee,* based on such criteria as the prevailing codes, feasibility, cost-effectiveness, financial benefit, health and safety needs, and optimizing the facilities and the work environment;
- identifying sources of energy financing and weighing the relative financial benefit for various financing options; soliciting performance contracting proposals and evaluating ESCO qualifications;
- serving as a liaison and contact point for the performance contractor;

- implementing approved recommendations: writing specifications as well as overseeing procurement, installation, fine tuning, and operation;
- planning and implementing internal and/or external communications strategies, or supplying energy information to those responsible for that function; and
- evaluating the energy program's effectiveness, updating it, and routinely reporting progress to top management and/or the board.

An effective Energy Manager needs to have technical expertise and some financial insights as well as communications and leadership skills. Unfortunately, persons with technical expertise do not always have the verbal skills required. The available choices range from a truly fine engineer with limited communications skills to the eloquent leader, who doesn't know a Btu from a cup of coffee.

Given a choice, someone with balance in technical and communications skills is desirable. The ultimate answer obviously varies depending on what is needed, what engineering or communications capabilities are already on staff, and the relative importance attached to hardware vs. user considerations. But if pressed to make a choice, take the leader and teach them the technology. Many energy engineers, bless their hearts, are great with equipment, but lousy with people. And let's face it, there may be no way that they can be groomed for effective leadership.

The Energy Manager must be in close and constant communication with the people who pay the bills. A brief session involving the financial officer, the Energy Manager, and the utility representative to assess demand profiles, pricing options, load management issues, etc. and their implications for operation can often reduce energy cost by thousands of dollars. This can, and should, be done before a performance contract is entered into; so these easily secured savings do not have to be shared with an ESCO. If it is not already done when the ESCO enters the process, then the ESCO should be asked to facilitate the process and be a party to it. Even then, the organization can, AND SHOULD, keep most of these savings earned through a better understanding of the purchasing process.

As volatility in the utility industry once more rears its ugly head, this utility consultation practice will need to be done more frequently. Someone, who has special expertise in supply acquisition, such as a power marketer, may be warranted.

ENERGY SECURITY

A new and growing responsibility for Energy Managers is assuring the availability of power to the organization. The energy security issues embrace health, safety and profitability. Since energy security is a relatively new concern, it has been singled out here to respond to that need and to serve as an illustrative example of how each EM responsibility needs to be developed.

Dr. Landis Kannberg, who has had extensive experience in security planning, reminds us that panic planning leads to serious errors and misunderstandings and often puts us squarely in the "paralyzing power of uncertainty." The following planning points are expanded in *Manual for Intelligent Energy Services.** As noted in that book, much of the guidance offered here draws on the excellent work of Dr. Kannberg.

1. Develop a security model based on mission & values, make it part of your culture and goals. The model should embody the risk posture of the organization,
2. Establish executive advocacy; clear, direct lines of authority for comprehensive security,
3. Treat security as a risk management issue. By including security of assets as part of the risk, security investment decisions gain the benefits provided by rigorous risk management approach, and reduces gaps in risk acceptance,
4. Know, and internally communicate, your critical assets and their security/risk management requirements,

---

*Shirley J. Hansen, *Manual for Intelligent Energy Services.* Published by The Fairmont Press, 2002.

5. Conduct vulnerability assessments periodically,
6. Review business continuity and emergency response plans and procedures, and periodically test them,
7. Examine your interdependencies,
8. Ensure appropriate trust allocation,
9. Examine security of partners, suppliers, contractors, even clients—join them in ensuring consistent security, and
10. Don't confuse reliability with security: both are important.

Whether you are going to serve your security needs internally, outsource some of them, or both, the elements of a security plan make a good case in point as to how the Energy Manager's job has grown and the responsibilities have spread. As conditions change, it's important to remember that a security plan is a dynamic document and needs to be revisited regularly as needs and personnel change

PROJECT ASSESSMENT

In the early days of performance contracting, the merits of "shared savings" used to be billed as "Their Cash; Your Savings." Like a complimentary dinner guest in a restaurant, the guest doesn't worry about the size of the check. Through the years, however, has come the realization: while the customer may not pay up front; at some point, *the customer will pay*. The funds may be "off balance sheet," money that previously went to the utility, but no matter how one slices it: ULTIMATELY IT COMES OUT OF THE OWNER'S POCKET.

As in all things, the customer should assess what the organization is getting for its money before signing a contract, during the installation phase and throughout the life of the contract. If the organization doesn't have the "how-to-do-it" resources in-house to make this assessment, they are available outside. The owner need not be impeded by the front-end cost of such services; for, unless they become too exotic, they can all be assigned to the project. Outside resources may include:

- an engineer with energy efficiency credentials (not an architect unless their training or experience includes mechanics, electricity and energy efficiency);

- a contract attorney with performance contracting experience;

- a financial consultant with an understanding of life cycle costing and energy efficiency;
- an independent assessment group to do periodic checks on the formula, the adjustments to it, and the savings calculations;
- measurement and verification (possibly by an impartial third party); and
- a performance contracting consultant, who can look out for the organization's best interests, guide it through the selection and implementation process, and put the organization in touch with top notch people who provide the above services.

Effective project management from the Energy Manager's side also demands a means of independently assessing the savings achieved over time. A computer-based energy accounting program, should be an integral part of an Energy Manager's project assessment practices.

Chapter 9
Communications Strategies

More performance contracts end up in court due to a lack of communication than any other reason.
—David Birr

The critical link between the ESCO and the owner throughout the life of the project is communication between the Project Manager, on behalf of the ESCO, and the Energy Manager, who represents the owner. Establishing communications channels between the Project Manager and the Energy Manager and preserving those connections are crucial to project success.

To elicit support from the building occupants, the Energy Manager must be able to articulate energy usage and benefits of efficiency in a fashion that strikes a responsive chord in the users. Generally, this means overcoming the frequent perception that "energy conservation" means achieving lower usage by giving up some quality of service or level of comfort. It can be helpful here to focus on *energy efficiency* as providing the same, or better, energy services and level of comfort while using less energy. Energy *conservation* by definition means using less—even at the expense of the work environment; so the term, "conservation," should be used judiciously.

Nevertheless, even with a declared emphasis on "efficiency," there remain stories such as the Energy Manager who was told by the ESCO, that in order to achieve the desired level of efficiency, the building must maintain a temperature of 82 degrees—even during summers of 90 degree heat! Despite his concerns, the Energy Manager thought he had no choice and ultimately occupants, Energy Manager and the ESCO were miserable. A major part of the Project Manager's job is to support the Energy Manager. The responsibility to implement strategies that will gain support from management and occupants should ultimately rest with the Energy Manager.

It is absolutely essential that the Energy Manager make that long, long climb from the control room to top management. Management's understanding and visible support can provide the Energy Manager with the authority and the clout to get the job done.

LINE OF AUTHORITY

Who should an Energy Manager report to? Experience has shown that in order to effect change, an Energy Manager must hold a position of stature within the organization's hierarchy. While situations will vary with personalities and local conditions, it is generally advisable to have the Energy Manager report directly to top management; e.g., the CEO, the hospital administrator, school business official, financial officer, superintendent or president. Some factors to consider in making this decision are:

- Decisions about operations (air exchanges and circulation containment needs, class schedules, athletic events, etc.) frequently necessitate discussions with directors, principals, department heads, deans, medical staff or sales managers, who all have their own worries and goals. Such discussions must be held among individuals with approximately the same organizational stature. It is hard to effect change from a position buried in the facilities or engineering department.

- Implementing an energy program almost always requires changes in past practices and procedures and it is not unusual to find resistance to these changes. If the "boss" is the facility manager and it is the boss' past practices that need changing, making a difference can be exceedingly difficult. Unfortunately, one does not have to look very far to find someone happy to follow behind and erase all the Energy Manager has worked to achieve.

If the Energy Manager is to "make a difference" in operations or maintenance, he or she must be on a parallel footing with the director of facilities, chief engineer, etc.—not reporting to him or her.

The relative position, real or conferred, that the Energy Manager holds in the organization is regarded as an expression of the emphasis management places on the energy program and its commitment to it.

An ESCO can learn a great deal about a potential customer and that organizations attention to energy by checking where the Energy Manager is positioned on the organizational chart, the level of authority and the budget allocation.

Effective Communications

The ideas and concepts considered in previous chapters and throughout the book may be quite new to many of the people whose cooperation is essential for energy efficiency and performance contracting success. Members of a management committee, a board of directors, and other managers must understand what the performance contract aims to accomplish, what procedures will be involved, and what the results are expected to be. Many "higher ups" will want to know how this benefits the company or helps the organization achieve its mission. Many, in other departments that will be involved, such as finance, will want to know why they should add on yet another responsibility.

Operations and maintenance staff must understand the "why" as well as the "what" of their role in energy efficiency, or the effort will fail. Care must also be exercised to assure O&M personnel that the new energy effort is not a criticism of their past efforts, and will, in fact, enhance their work.

Building occupants need to know how they fit into the efficiency plan—and how they, in turn, will benefit. Without that understanding and the commitment to help, occupants can defeat the most sophisticated control system or the most beautifully articulated management plan.

Once management starts the process of seeking an energy service company partner, communication becomes an essential component of the selection process. Once the contract is signed, regular progress meetings and close coordination become essential components of an effective project.

There is nothing magical, nor particularly sophisticated, about the application of communication techniques to the various requirements of an energy efficiency program, or the performance contracting aspect of it. Effective managers know very well how to communicate with their boards, staff, clients, patients, customers or community. The absolute dependence of an energy efficiency program on the human element, however, makes a review of communication fundamentals valuable. An energy efficiency project spends money on hardware, but whether energy and money is actually saved depends on people... and getting results depends on effective communications with the people involved.

ELEMENTS OF EFFECTIVE COMMUNICATIONS

Most aspects of an effective ESCO's operation are common to any well-run organization; however, energy service companies have a few peculiarities that require special management sensitivities and procedures.

First, ESCOs sell a service, but customers usually think they are buying products. Performance contractors sell cost-effective productive environments; the boilers, chillers and controls that may be installed are merely vehicles to make it happen. The hospital needs a new boiler; from the administration's perspective, getting a more efficient boiler through performance contracting, means the boiler is paid for from future energy savings. This difference in point of view can undermine effective communication between the two organizations and frustrate the sense of partnership. Understanding where the other guy is coming from is key to effective cooperation and successful projects.

Second, the ESCO is selling promises; predictions of savings. When those savings are realized everyone is happy. But memories are short. The high utility bills of the past may be forgotten, especially if there has been a change in management, and soon the customer may feel he's paying for "nothing."

One important ingredient ESCOs can, and should, provide their customers is cost avoidance information. A performance contract is designed to save ENERGY; money savings are an important by-product. However, performance contracts sold as "money savers" can become management headaches if the price of fuel goes up. ESCOs, who want to develop and maintain good relationships with their customers, learn how to calculate cost avoidance and to communicate those "savings" graphically with billing procedures, through regular briefings, annual reports, etc. Information on avoided costs are a vital part of a communications effort. (See Chapter 3 for a description of cost avoidance.)

Third, to the uninitiated, the process may seem to rely on smoke and mirrors. It requires some expertise in both financing and energy technology if the customer is to be comfortable with the performance contracting concept. Bridging the gap from business office to boiler room is not always easy, but can be done if the ESCO and the customer work together effectively.

Empty Conduits or Real Substance

The two real issues today that bolster the need to do energy efficiency work are *"environment"* and *"money."*

Environment—there is growing concern about the pollutants emitted by burning fossil fuels. The question of the 1970s was "do we have enough fossil fuels to sustain the economy?" The question for the 21st Century is, "Can we afford to burn what we have?" That question is of global concern, particularly as we consider the growing economy of developing countries, such as China, whose energy demand is increasingly exponentially.

According to Joseph Romm, in *Lean and Clean Management,* "Efficiency lets a company do very well while doing much good, reducing emissions of nitrogen oxide, sulfur dioxide, and carbon dioxide—gases that cause smog, acid rain, and global warming."

While "environmental issues" will drive many individuals and organization to become involved in energy efficiency, it is well to remember that energy costs drive the economic viability of a project.

We need to talk ***money***... dollars saved rather than energy conserved.

Everyone understands and is interested in money. Few would agree to throwing money out with the garbage; so you just might interest them in what they are burning up. Even though a performance contract is written in terms of energy saved (Btu or kilowatts) talking money gets the message across.

Cultivating and making use of that interest in relation to energy efficiency is the communicator's primary challenge. In fact, if those dollars can be translated into a competitive advantage, enhanced bottom line, computers, band uniforms, etc.—something of great interest to members of the organization—THEN "energy savings" become even more attractive. Look at the people from which you need to gain support, consider their individual interests; then, use that information to move forward and win them over.

A few years ago, Mr. Anil Akuja, former energy program manager for the Los Angeles United School District, described an effective communication's strategy used to gain support for a $15 million energy program. He said the new energy management program appealed to different decision-makers for different reasons:

> We sold [the concept of installing an energy management system] to the schools' principals by saying the systems would improve the classroom environment. The board bought it because the systems would have a 3.5-year payback, and the maintenance people bought it because with the alarm functions of the systems, their operations costs are lowered.

Understanding the art of communicating to the needs of specific groups, Akuja concluded that the program received support from all board members because it had saved the district $67 million in avoided energy costs.

Today, future energy costs remain an unknown. Most believe that prices will trend upwards, some believe there will be power outages, and some believe customers will be demanding more services for their money. Nearly all agree there is uncertainty. What better reason to save money than the uncertainty of the future!

COMMUNICATING ENERGY NEEDS

Effective energy management means that Energy Managers and Project Managers must understand and fully accept the critical role of communications in all phases of the energy efficiency effort. One attorney involved in a performance contract arbitration commented "nearly all the problems seem to be traceable to in-effective communications."

Effective communications is *planned and orchestrated*. It is not *ad hoc*. The "Oh sure, I'll tell him when I see him" approach does not do the job. Done right, the tools of communications become as much a part of a successful program as does the work accomplished with a wrench or a screwdriver. The man knew what he was talking about when he said, "The pen is mightier than the sword."

The essence of effective communications lies in knowing: (1) the target audience... *Who* do you need to reach? (2) the purpose of the message... *What* action do you want to prompt? (3) *What* (and how much) do they need to know to achieve the results you want; and (4) *What* is the best time, route and format to use to reach your audience. How well these factors are understood and how they relate to each other will determine the degree of success (or failure) of any communication's effort.

Target audiences can, and must, be defined. At a minimum, they include top management, staff, and building occupants. People in the community are a particularly important group to public institutions. If an organization is to function smoothly and effectively, none of these audiences can be neglected for long. And each requires a message tailored to that group.

BRINGING THE "BOARD" ON BOARD

The "blame game" can be easily traced to boards, which do not have the written, visible commitment that a quality energy efficiency project requires. No energy efficiency project can be truly effective without solid

commitment; therefore, it makes sense to start this discussion of communications strategies with the board, the governing body or top management. Not all organizations have "boards"; however, for simplicity's sake, the term board will be used in a generic sense to encompass all governing bodies and the higher echelons throughout this discussion.

The board needs to be informed regarding past energy management successes in similar organizations or companies. Information about what money can be saved can be crucial. If any part of that understanding is incomplete, getting backing for new energy projects can be very difficult.

A few facts can help tell the story. Does the board know what energy costs are for each unit of production? Per pupil? Per square foot? Per bed? Per widget? Do they have available information about the potential to save? Are utility matters a real part of the budget process, or is it just another line item that can be passed over without discussion because "nothing can be done about it"? How well have energy security issues been addressed?

A simple presentation to the board showing the money the organization is losing unnecessarily and the dollars that might be saved through improved energy efficiency is a good starting point. Try out some hypothetical (but reasonable) numbers, say 15 percent or 20 percent savings this year, and then project those savings for 5 or 10 years. Then, ask for support to explore ways to make those savings possible. (Selling energy efficiency and performance contracting to the board is treated more fully in Chapter 1.)

Does the board know the environmental damage being caused by the unnecessary burning of fossil fuels? The CO_2, NO_x and SO_2 emissions do not have to be calculated by each fuel exactly—some figures averaged to Btu reduction will get the message across.

The board also needs to hear about the potential for energy interruptions due to possible terrorist activities and the plans that have been formulated to address such needs. If you are a designated first responder hospital and you are considering stationary fuel cells, now is the time to bring it up.

A walk-through energy audit of several buildings will provide a good estimate of savings opportunities and suggest the steps needed to get the job done. This should also supply the basis for determining the dollar and environmental costs of doing nothing about the problem. At this point, we are not talking about a full scale audit!

"Energy" information needs to be presented to the board in a straight forward manner. Burying board [or bored] members under discussions of Btu or elaborate discussions of variable air volume will only confuse the issue. Also, it's important that the discussion not become a criticism of present or past management of energy resources. A general discussion of what can be done, the dollars that can be saved, and what those dollars could mean to the operation (supported by environmental concerns) should be the basic message.

On almost any board, or in any management group, there will be one or more members who want lots of details about almost everything. In energy, as in other areas, it pays to deliver those details, but only on request. Delivering supplementary materials or having additional discussion outside of board meetings may recruit some solid support, while not turning off those who really don't want to know that much about air intake volume, burner efficiency, or CO_2 emissions.

Every CEO, president, superintendent, financial officer or chief administrator knows how quickly board members learn that anything anyone wants to do costs money. At this stage, it is valuable to introduce the idea that there are a variety of energy financing options available beyond traditional methods—and that several of these options do not require any capital investment, or "new" money. Support to explore those options and develop a plan for their consideration comes next. Keeping up the flow of information and moving ahead steadily, but in rather well defined steps, will enhance the chance of success when it becomes time to "ask for the order."

Performance contracting differs rather widely from traditional financing methods. As every administrator knows, anything out of the ordinary, particularly when it comes to financing, often makes board members uneasy. As noted in Chapter 1, it is important to stress that energy efficiency can be self-funded, and therefore, only requires the redirection of money already in the budget, presently going to the utility for wasted energy. The ability to explain the options that are available with considerable clarity and confidence takes on added importance. An outside consultant may provide valuable guidance and communications support.

Citing others who have successfully used performance contracting may allay concerns. The fact that both the public and private sectors are using performance contracting with increasing enthusiasm speaks to its effectiveness.

It is also desirable to remind the board and top management that energy efficiency is an investment not an expense and that these *investment* opportunities can be self-funded from money now going to the utility for wasted energy. Let them see the opportunity: money now paying for wasted energy can be invested AND they can get as much as 50 percent return on investment for that money!

OUTSIDE AUDIENCES

Talking to external publics can be very important, especially for public entities. For a public entity, the community is the ultimate boss, paying the taxes, and voting for council members. It pays dividends to let the community know that the administration is doing something about energy costs and the environment. Audits and studies aimed at increased efficiency are news. Training programs to upgrade operations and maintenance skills are news when couched in terms of controlling energy costs and pollution while improving the learning, patient care, or work environment. Stories can be prepared for the local press, or additional opportunities to get the word out can be found.

When the time comes to move forward with a performance contracting project, the information about controlling energy costs and pollution without using local or state tax dollars is good business and good news for local governments, schools, etc.—and the tax payers! Proceeding in a businesslike manner to do something positive about costs, improving cash flow and freeing up funds for special purposes without adding to the tax burden, is very welcome news indeed.

When an RFQ/RFP is issued, public organizations should let their external audience know about it. It reinforces the fact that something is being done to use resources more efficiently. When contracts are signed, publicize the fact that a program is off and running. If it is appropriate, emphasize the portion of the work that will be accomplished by local subcontractors. Work for local companies is always good news in the community. If the project is being financed by an energy service company, it helps to coordinate the development of any news releases with them to assure consistency and accuracy.

Talking to an external public can also be good business for an industrial facility or the operators of commercial buildings. Let people know the organization is a good corporate citizen, saving resources and helping to protect the environment. "Good news" about what you are doing helps with community, and employee, relations.

ESCOs can be very helpful in supplying public information strategies that have worked with similar customers. A performance contractor can become the conduit to networking with other organizations with performance contracts. The beginning of work and the installation of new equipment can provide a "news hook" ... an excuse ... for communications to all who are interested that things are happening on the energy/dollar front. It provides still another chance to build support for, and an understanding of, the energy efficiency program.

Consider photo opportunities. If the energy savings "bought" new band uniforms, show the students trying them on. If the savings paid for new computers for the college, show them being unloaded. If the savings paid for replanting the atrium at the nursing home, show the planting with residents helping or looking on. If the savings increased the donation your firm can give to the Salvation Army, get a picture of presenting the check—or better yet show the good things that were made possible with the money.

BRING THE STAFF ON BOARD

Without question, successful communications with staff is vital to the mechanical and financial success of any energy efficiency project. If the staff, at all levels, understands how the energy effort will benefit *them*, they will generally support the program. If not, determined "resisters" can find ways to defeat any plan—consciously or otherwise.

It is important to use every opportunity to put across the message that the energy efficiency effort is not just to save energy but to save dollars that could be going for other, much more desirable things.

That is why one portion of an organization's needs assessment should be devoted to developing a "wish list" and calculating ways to show how energy savings can pave the way to getting computer systems, books or new equipment ... or cutting product/unit costs. While freed-up dollars might not be immediately available to purchase the items on the wish list, seeing the opportunity to do so will make the benefits of the savings more real for many.

Building occupants also need to know that energy efficiency doesn't mean bundling up and freezing during the winter, nor does it mean living with far too much heat in the summer. We are still haunted by the Emergency Temperature Building Restrictions of the late 1970s that were part of the Emergency Energy CONSERVATION Act. Since then "conservation" has been equated by many to deprivation. It is good business to use the term energy *efficiency*. After all it is what we are really about: using the energy that must be used for a safe, comfortable, productive workplace as efficiently as possible. A well engineered and executed energy management program can enhance overall comfort levels. Of course, some buildings will never be totally comfortable no matter what is done to the HVAC system; so it pays to be cautious with any promises in those situations.

And don't forget that the environmental message is just as important to the staff—maybe more so.

The total building population is an "audience" that will experience the results of the program. If heating and cooling systems work better (or worse), what is perceived to be happening will make all the difference in their acceptance of changes that occur. If this audience knows and understands what is going on and feels that the results will benefit each of them respectively, fewer instances will be created where, for example, an instructor, aided by students, destroyed a thermostat with the heel of a shoe. Communications to this vital inside audience can be tailored to fit various segments of building populations, so everyone can be let in on the "secrets" of energy efficiency.

As an important side benefit, it has been demonstrated over and over that effective, internal communications is one of the most effective external communications tools. Nurses talk to patients, staff to friends, students to parents and to other students, plant workers to spouses and the word spreads.

Operations and Maintenance

A critical first step in communication with operations and maintenance (O&M) staff is to affirm the good things they have been doing. They must not view preparations for an energy efficiency program, or a performance contract, as a criticism of their work. Nor should it be viewed as a job threat. Performance contracting can free the staff to do other work that has been put off far too long. (Unless, in fact, the plan is to secure some savings from reducing personnel, which is a decision that should be made TOTALLY by the customer.)

O&M personnel should be active participants in the effort from the beginning and should receive a steady flow of information as the project moves forward. Even when the process becomes a matter of RFQs and contract writing, the O&M staff should be kept informed of progress. They have a direct, personal interest, because when the project actually "happens," it will happen to them. Their attitude will have a major impact on results—good or bad.

GETTING THE WORD OUT

Messages need to be kept simple. Most people are not interested in all the details of the project, but they do want to know how it affects them. A few thoughts for guidance include:

- Tell building occupants what it all means in terms of comfort and money saved for things of interest to them;
- Tell O&M staff what the changes can mean in improved equipment to work with, perhaps fewer complaints from occupants, better levels of maintenance and control.
- Tell employees about comfort as well as dollars—and what they can buy from the money salvaged from energy bills.
- Tell the board (and the public) that there are ways to increase energy efficiency, gain capital equipment, reduce environmental damage and increase cash flow without the commitment of organizational funds. Those who want all of the details will ask and should be accommodated, but for the most part the real question to be answered is, "What does it mean for me?"

Communications are a key part of energy efficiency planning and should be considered a part of every phase. When the audits are done, publicize the fact, giving credit to those employees who do the work. When consumption drops in particular facilities, recognize the custodians or key maintenance personnel at a board or management meeting. An investment grade audit report is a guide to effective investments by the owner; it's also a guide for informing the public of the investments that are being made and why.

Projects do not always go smoothly nor precisely follow the plan. Careful attention to communications throughout every phase provides the needed base of credibility if things should go wrong. And if they do, acknowledge problems while they are small and establish the fact that steps are being taken to deal with them. That way, there can be no sudden revelations of CRISIS. The keys to successful passage through the communications mine fields when problems arise are openness, candor, and adherence to the first (and best) rule of press relations: NEVER SPECULATE.

If problems arise, never cut off the flow of information.

GETTING THE JOB DONE

Who should do all of the communications work? And, it is work. If the organization has an information director or public relations person, this effort should be a real part of his or her responsibility; but, it cannot be done alone. There must be a steady flow of information from those directly involved in the project to the person responsible for dissemination. This type of information effort often fails because someone is "too busy" to pass along details of what is happening and / or doesn't realize its communications importance. The information program will be effective only if it is considered as a regular part of the project and planned as carefully as engineering and finance. If communication is handled as an after thought, or "when I get around to it," there is a real risk of project failure.

The day when everything is finally in place and running well is a good time to celebrate. It is time to get the word out yet again. A time to review what has been done and to look ahead to what all the changes can accomplish ... a time to "point with pride." It is also an opportunity to salute the staff and recognize those who have helped make it happen. Recognition ceremonies may be appropriate. More than one certificate has found its way to the boiler room wall. Also express appreciation to those who may have been inconvenienced during the project. It's a great time to say thanks.

And when you have results, SHOW THEM. Keep the picture opportunities in mind. If management can demonstrate that the energy efficiency program has saved money for those special items from the "wish list," tell the story. Installing new lab equipment, patient care equipment or new playground equipment in the city park, which were paid for through energy savings, has real visual impact. This is great material for company employee newsletters or magazines.

Communications strategies are basically the application of common sense to the distribution of information. Successful managers have learned to use the tools and techniques of communications in all sorts of circumstances. An energy efficiency program depends upon the application of the best of those skills if it is to get results. With a performance contract, the partnership is strengthened through planned communication.

Section II
Managing Risks

It seems self-evident that an industry, which depends on guarantees, must carefully consider risks and the way they are managed. It cannot be overstated: performance contracting is risk management!

The ability to realistically assess and manage the risks ESCOs encounter is the foundation of an effective project for the ESCO—*and the customer*. Effective risk management is the most predictable measure of an ESCO's future success.

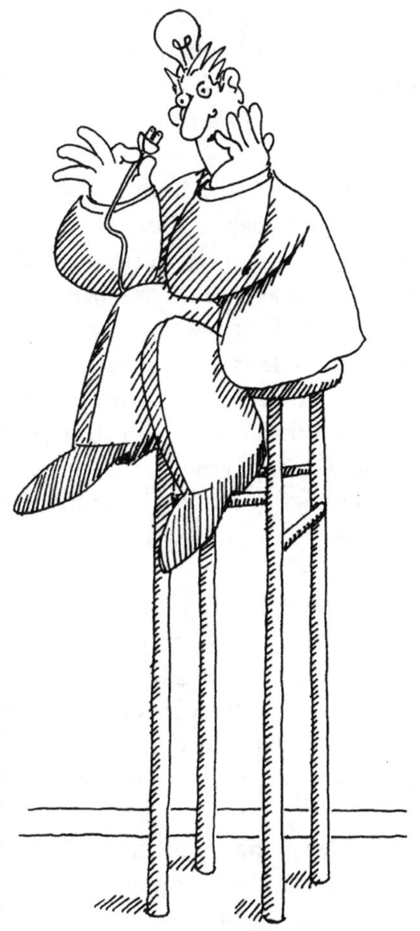

The whole idea of "risk management" prompts a rather frightening specter for some; so it is probably prudent to provide a brief overview of risks, their identification and management before discussing the particulars. We tend to put "risks" off in a special compartment with a large "Do not disturb" sign, but the truth is we practice risk management every day from the moment we stick our big toe out from under the covers each morning. It's as simple and reflexive as looking both ways before we cross the street.

Risk management starts with identifying the risk and evaluating the source of the information. Have you heard, for example, that coffee is bad for you? What was the source of that information? Was it reliable? Did you stop drinking coffee? For years many of us have avoided coffee or substituted decaf or tea; now the "experts" tell us it is good for us! How do we manage the coffee "risk" today?

Once the risk is identified, an assessment should be made to weigh the potential risk against the expected benefits. Part of the assessment is a careful examination of the variables than can exacerbate or diminish the risk. If it is determined that the benefits outweigh the risk (or the cost of mitigating the risks is outweighed by the benefits); then some procedure to manage the risk and its associated variables is engaged.

In every case the management of a risk involves decisions regarding:

a) the relative risk/benefit ratio;
b) whether the risk, or its associated variables, can be reduced or eliminated cost-effectively;
c) the level of risk that is acceptable; and
d) who can best control and bear the risk burden.

We can draw from the insurance industry to examine the range of risk management options. The primary internal management strategy to control the risk is through substitution, reduction or avoidance. Consideration of these options can focus on the risk or on the associated variables. The selected management option(s) will vary with the nature of the risk itself; the relative cost of a specific mitigating strategy; and its impact, if any, on the benefits to be achieved.

Another option, which is particularly effective in energy efficiency matters, is assignment. An example, is the energy floor price used in the cost savings guarantee in guaranteed savings model.

Assigning of risks is based on:

a) who will accept the risk,
b) who can BEST accept the risk;
c) what, if any, mitigating strategies can be implemented;
d) what are the concerns associated with the various in-house strategies available; and
e) what are the costs of outsourcing the risk or its associated variables.

Two axioms prevail: (1) risk is typically best assigned to the party that can control it or gain the most from its control; and (2) the benefits from engaging in an activity must outweigh the costs of mitigating/managing the associated risks.

Risks may be real or perceived. If a key decision-maker in an organization perceives a risk as real, then, for all practical purposes, it is real—and should be treated as such. It may, however be possible to convince the decision-maker that it is a perceived risk. A softer option may be to treat a potential variable of the perceived risk; thus having "treated" the risk.

Whether risks are real or perceived, they can affect the financing available. Risk factors can constrain the procedural options, limit the availability of financing, or determine the cost of money. Risk, and strategies to manage risks, always carry a price for someone. *Money always follows risk.*

It is well to remember that due diligence of financiers in considering the merits of a proposed project is itself a safeguard and serves as a second opinion as to the risks that may be incurred.

Owners should also understand that failing to act may carry its own risks—fiscally, professionally and politically. A school district on the east coast was laying off uncertified personnel due to budget constraints; yet, at the same time, wasting over an estimated $100,000 per month because it was slow to implement a performance contract. When the union learned of this negligence, the consequences for non action became serious.

The best course of action is to judge the alternatives prudently and select the course that offers the greatest, or most critical, benefits for an acceptable level of risk. This section offers owners and ESCOs some guidelines in assessing and managing a range of risks related to energy management. A particularly important portion of sChapter 11 is the discussion of how the financial structure is used to help manage risks. It clearly illustrates how much the owner as well as the ESCO has to gain from working closely with the project partner to minimize project risks.

As a case in point, Chapter 12 discusses one of the greatest risks facing owners and ESCOs alike: indoor air quality (IAQ). Energy efficiency and IAQ are inextricably linked through perception and reality. Misinformation abounds. Energy efficiency does not necessarily constitute an IAQ threat; in fact, as noted in Chapter 12, there are many opportunities to have energy efficiency and IAQ together effectively. And to benefit both areas!

Chapter 10

Owner Risks and Mitigating Strategies

Owners need to weigh the relative risks associated with each of the performance contracting delivery options within the context of local conditions. This chapter defines the major risks and discusses the variables that exacerbate or diminish those risks. It is designed to help potential performance contracting customers decide whether to bear or transfer certain risks as part of the project delivery strategy. If in-house staff cannot mitigate the particular risks associated with the selected option(s), then an organization may choose to transfer these risks either by selecting a project delivery option that assigns these risks to the service provider or outsourcing directly for services that will mitigate the risks.

Risk factors will vary with local conditions at particular times. Effective assessment and management procedures need to reflect these local conditions. Sometimes we find an organization is its own worst enemy, and portions of an operation simply defeat the efforts of another. Such risks, which seem peripheral to performance contracting, may still constitute a risk burden for the project.

Once adjusted for the unique local characteristics, the costs associated with accepting or assigning risks can be determined. These "costs" go beyond money and frequently encompass time, manpower, administrative capital and/or political considerations.

In many instances, there are several management/mitigating strategies available. Weighing the price tag and other costs for each strategy should be a cooperative effort between the owner and ESCO.

The Risk Analysis Framework

Performance contracting can offer organizations risk shedding opportunities. Evaluating various performance contracting structures requires an understanding of the relative risks each carries. The op-

tions examined are vendor financing, shared savings, and guaranteed savings.

The range of out-sourcing opportunities addressed within this framework is those unique to performance contracting. For ease of reference, the risk management frameworks presented in this chapter separate the technical, financial and procedural risks.

The following analysis was first developed at the request of Ms.

Christine Vance, Bureau of Energy Conservation, City and County of San Francisco. This work was a part of an effective energy efficiency financing decision making model, *Picking Up the Pace*, which the Bureau developed under a grant from the US Department of Energy's Urban Consortium. Ms. Vance and her associates in the City of San Francisco are to be commended for realizing that potential performance contracting customers, particularly those in the public sector, need a realistic way to identify risks as well as ways to compare and mitigate such risks. The authors gratefully acknowledge the customer perspective provided by Ms. Vance, her colleagues and associates, which is incorporated within the following discussion.

PERFORMANCE CONTRACTING OPTIONS

Before the risks related to various performance contracting/financing options are assessed, a quick summary of the energy efficiency financing options available through performance contracting will make the following figures more understandable.

Vendor financing. Typically, the simplest form of performance contracting, vendor financing, is generally offered by a manufacturer who wishes to demonstrate confidence in the energy efficiency capabilities of its equipment and offers to take payment for the equipment out of the avoided utility costs. This approach is often referred to as "paid-from-savings." Financing (and equipment selection) is limited to those vendors offering such a service. Bias by the vendor towards his own equipment is a factor.

Shared savings. Prior to project implementation, the owner and energy service company (ESCO) agree on a percentage split of the energy *cost* savings. Performance and credit risk are both carried by the ESCO. If there are no cost savings, the ESCO does not get paid. If savings are greater than expected or energy prices go up, the customer can easily pay more than expected for the use of the equipment.

Guaranteed savings. The ESCO guarantees the quantity of energy to be saved and that the dollar value of those energy savings will be sufficient to cover the debt service obligations. The customer incurs a credit risk on the books, but the debt service obligation is guaranteed to be met through the savings unless the price of energy drops below a specified floor price.

TECHNICAL RISKS

The major risks typically associated with energy efficiency work are technical considerations, which run the gamut from equipment selection to the technical expertise of in-house staff and outside consultants, as well as energy audit quality, construction/installation matters, maintenance and operations concerns, savings persistence potential, measurement and savings verification. Each of the these risks may vary with the financing mechanism used. These are presented in Table 10-1, with the relative risk associated with the typical energy efficiency financing options, the major contributing variables to the risk, as well as potential mitigating strategies to treat those variables.

Many technical risks can be managed through an analysis of the variables contributing to the level of risk and implementation the appropriate mitigating strategies. All mitigating strategies, however, demand internal resources and/or the expense of outside support. In almost all cases, for example, direct purchase will reveal the highest level of risk to the organization, but may require the lowest expenditure.

Of the three types of performance contracting presented in Table 10-1; i.e., vendor financing, ESCO shared savings, and ESCO guaranteed savings, the guaranteed financing approach generally offers the greatest level of risk shedding opportunities. Vendor financing is usually equipment specific; seldom offers a comprehensive energy management approach; and is constrained by the vendor's line of equipment, probable equipment bias, and potential "needs" enhancement.

Vendor financing and ESCO shared savings offer off-balance sheet financing, which may make them more attractive when weighing certain financial risks or where debt ceilings are a factor. Neither approach offers the comprehensive technical support and the risk shedding offered by ESCO guaranteed savings.

Since an ESCO carries both the credit and performance risks in shared savings, the cost of money is higher. This, in turn, drives down the acceptable payback period and frequently removes the big ticket items, such as boilers and insulation, from projects. To gain the benefits of guaranteed savings, particularly technical risk shedding, the owner must accept the credit risk and the associated costs.

Table 10-1. Technical Risk Framework

TECHNICAL RISKS	PERFORMANCE CONTRACTING OPTIONS			MAJOR VARIABLES	MITIGATING STRATEGIES
	Vendor Financing	Shared Savings	Guaranteed Savings		
Equipment performance - longevity - warranty	3	4	1	Quality of specs Selection process Contract conditions	In-house/consultant expertise Legal ability available
Technical expertise	3	1	1	In-house staff experience/training Consultant qualifications	Provide needed experience/training Secure outside consultation Selection process
Audit quality; accuracy	4	3	2	Auditor technical and risk assessment abilities Review capability Vendor bias	Selection process Review a sample audit 3rd party validation Establish procedural criteria & scope of audit
Construction/ installation	1	2	1	Vendor or subcontractor qualifications Contract provisions	Selection process Performance/payment bonds Legal ability Owner construction supervision
Maintenance & operations	3	3	1	Manpower In-house staff qualification Training quality	Outsourcing Training; experience Selection process
Savings persistence	3	3	1	Varies by measure Administrative commitment O&M attitudes, training, experience	Contractual obligations Vendor selection ESCO selection Guarantees offered Deal's financial structure
Savings verification - approach - instruments	5	4	1	Needs vary by measure guarantee, & needed accuracy	Amount paid for accuracy 3rd party validation

Legend: N/A not applicable; 1 low risk; 2 low-medium; 3 medium; 4 medium-high; 5 high

FINANCIAL RISKS

Financial risk factors are paramount in evaluating the best procedures for funding energy efficiency measures. Risks associated with various energy efficiency financing approaches are typically significant factors; including securing the financing, impact on the debt ceiling, the likelihood of achieving the savings, the cost of money (interest rates; use of tax-exempt status, and repayment terms), energy price fluctuations, and the extent of equipment warranties. These risks also include other important considerations, such as whether the guarantor will be there to back the guarantee for the life of the project, what hidden costs may exist in package deals, and the higher payment risks specifically related to shared savings.

Table 10-2, presents these risk factors and the typical level of risk associated with the factor by financing option, major variables and associated mitigating strategies. In most cases, the risks for all three performance contract options can be managed through contract provisions. In all cases, the customer's willingness to assume performance risks will typically lower project financing costs.

Before discussing the specific implications of the risk factors revealed in Table 10-2, one other factor needs to be addressed. Since the financing is carried by the ESCO in vendor financing and shared savings, there is a tendency to assume that the owner will not pay any transaction costs or finance charges. All costs incurred by the vendor or the ESCO must be accounted for somewhere in the project financing. While the financing typically comes from a larger pool of funds, the transaction costs are apt to be lower; however, the customer still pays for this cost of doing business.

Warranties

Like direct purchase, vendor financing is apt to have the typical manufacturing warranty and nothing else. However, the vendor's interest in getting paid from savings will help assure savings occur—and that the equipment will work until the equipment is paid off.

In shared savings and guaranteed savings, the ESCO payment depends on the equipment performing at a certain level for the life of the contract. This assurance is even stronger with guaranteed savings. So in addition to the manufacturer's warranty, which is typically conveyed to the customer under performance contracting, an implied warranty exists.

The ESCO guarantor could disappear, leaving the customer without this additional "warranty" and leave the owner with the debt service obli-

Table 10-2. Financial Risk Framework; Technical Issues

FINANCIAL RISKS FACTORS	PERFORMANCE CONTRACTING OPTIONS			MAJOR VARIABLES	MITIGATING STRATEGIES
	Vendor Financing	Shared Savings	Guaranteed Savings		
Projected savings not realized - warranty	1	1	1	Lack of expertise -specs -selection Vendor claims	Use outside consultant
Warranties -limited	2	2	1	Contract provisions Equipment selection	Negotiation
-manufacturer -guarantor "disappears"	N/A	2	3	Contract provisions Assume debt service obligation	Performance bond Organization's financial status
Increased savings or energy prices may make equipment payment too high	2	4	N/A	Contract provisions	Be aware Negotiations Set payment ceiling
Post-contract savings	2	2	2	Quality of the specs Project scope Equipment useful life	Engineering support Selection process Maintenance quality
Establishing baseyear; baseline adjustment provisions	4	4	4	Procedures Accuracy	Instruments & approach used Cost 3rd party validation
Fixed payment	1	3	3	Monthly savings fluctuations	Clear formula Vendor established
Cost of delay	1	3	3		

Legend: N/A not applicable; 1 low risk; 2 low-medium; 3 medium; 4 medium-high; 5 high

gations. This risk is underscored by the large vendor/ESCOs; e.g., Johnson Controls, Honeywell, Siemens Building Technologies, who point to their size and longevity as a positive factor. These big companies, however, are not without problems. In the rapidly changing corporate world, company size does not guarantee longevity. All ESCOs should be investigated as to their organizational and financial stability and when in doubt, a performance bond to back up payments should be required. Keep in mind, however, that the cost of such bonds come out of the project; so a share of the cost will be borne by the customer.

Increased Savings Or Energy Price Changes

In a shared savings project, the customer agrees to pay a percentage of the avoided utility costs to the ESCO. In a $1,000,000 per year savings deal, this might be 70 percent, or $700,000 per annum, which the customer could deem to be reasonable cost for the equipment and services received. If the avoided utility costs go higher than expected by virtue of greater than expected savings or higher energy prices, the customer may find itself paying 70% of $2,000,000, or $1,400,000. In a five year contract, where equipment has been valued at $3.5 million, the customer could find itself paying $7 million in such circumstances, double the equipment value.

To a lesser extent, this is also possible in vendor financing if the contract stipulates a term rather than just paying off the equipment costs. This concern can conceivably happen to some extent in guaranteed savings if the excess savings is shared with the ESCO. An ESCO's interest in helping the customer secure excess savings generally outweighs this problem.

Whether the selected financing is shared savings or guaranteed savings, this risk can be managed by stipulating the absolute dollar value of the shared or excess savings (a ceiling) the customer will pay.

Post Contract Savings

The length of time the installed equipment will continue to operate near design and, thus, deliver the desired savings is key when weighing project benefits. This risk can be effectively managed in any financing scheme by stipulating useful equipment life, quality of maintenance and performance criteria in the specifications and/or contract.

Fixed Payment

Not knowing from month to month what the required payment will be can be unsettling and can be a problem inherent in vendor financing or shared savings options. In fact, these fluctuations in shared savings have frequently created an adversarial relationship between the customer and ESCO to the detriment of the project. The bookkeeping by all parties is a greater burden. Fixed payments with annual reconciliation, which can be structured to work in shared savings as well, are preferable.

Cost of Delay

Small direct purchase items can usually be acquired quickly and the risks associated with the cost of delay would, consequently, be very low. For larger, more costly measures, performance contracting generally offers a quicker start-up—provided the organization's personnel and technical/legal support understand the concept. Increasingly complex measures frequently incur greater delays—and typically greater risks.

Intrusion/Interruption

The risks of interrupting operations or intruding on processes and procedures are typically higher in more comprehensive programs; thus, ESCO guaranteed savings are inclined to carry a higher risk and vendor financing a lower one. Requiring contractors to work outside of regular operating hours is a frequent solution, but the added labor costs are often borne by the project (and shared by owner) and must be weighed against the benefits.

Financing Availability

The creditworthiness of an organization is a factor underlying all financing mechanisms. The repercussions from Orange County, California, which surprised the non-profit financial community in the 1990s by declaring bankruptcy, are still felt in the non-profit market place. Cities and counties, for example, are subjected to greater due diligence and viewed with more caution than previously.

Shared savings options will be affected by the creditworthiness of the ESCO and its ability to assume any further debt equity.

Debt Ceiling

If a public sector organization is limited by statute, bond ratings, lack of voter authority or fiscal prudence from assuming more debt, vendor financing and shared savings, typically off balance sheet, are, therefore, more attractive options. If the public sector organization has limited room left to incur debt, the financing options need to be weighed against other needs, which may create demands on the credit available. Similarly, private sector firms must also weigh increased debt, for it can affect the ability to borrow for other purposes.

An organization pays in other ways for the off balance sheet opportunities; vendor financing is usually confined to a specific piece of equipment and does not offer a comprehensive approach. The organization is also limited to the product lines available from vendors which offer "paid-from-savings." Relying on vendor recommendations may also allow the "fox to design the hen-house." With regard to shared savings, the cost of money is nearly always higher; so the organization will get less project for the level of investment. Money drained off to cover interest also limits the size of the project—and profits—for the ESCO.

MONEY ISSUES

The ESCO carries both the credit and performance risks in shared savings; therefore, the ESCO must hold back funds to cover more risks. Financiers remember the number of shared savings projects that went sour in 1986, when energy prices dropped, and view a shared savings approach as a greater risk. In shared savings, the ESCO and financier are, in effect, betting on the future price of energy. With price volatility very much on the horizon, this risk has become even greater. Therefore, as depicted in Table 10-3, the cost of money for shared savings is high and is apt to go higher.

Cost of Money

Interest rates are always a factor in determining the most attractive financing option(s). In the United States, tax-exempt financing offers the most attractive interest rates for eligible institutions, but can only be used if a municipality, or other tax-exempt organization, accepts the credit risk. Vendor financing AND shared savings rely on the manufacturer/ESCO, which requires commercial interest rates. The amount of interest paid will depend on a number of factors, including how the deal is structured, the length of the contract, discount rates, and related issues such as the surety offered by large ESCOs.

In calculating energy payback periods, or return on investment, the investment is divided by total expected energy savings per year. Seldom, if ever, does the owner consider the reduced dollar value of those ener-

Table 10-3. Financial Risk Framework; Money Issues

FINANCIAL RISKS FACTORS	PERFORMANCE CONTRACTING OPTIONS			MAJOR VARIABLES	MITIGATING STRATEGIES
	Vendor Financing	Shared Savings	Guaranteed Savings		
Financing not available	N/A	4	1	Customer creditworthiness Performance risk of financing option	Use vendor financing or shared savings ESCO qualification and surety
Increase debt ceiling	N/A	N/A	5	Statutory ceiling Voter authority problems	Use vendor or shared savings
Cost of money - interest rates - discount rates	2	4	1	How deal is structured Length of contract	Contract conditions Financier opinion ESCO surety
Tax-exempt not Available	4	4	N/A		
Energy prices - fluctuation	3	4	2	Contract provisions Utility restructuring	Negotiation Follow state & federal actions
- negotiate rates; - prices may fall lower than floor price	N/A	4	4	Contract provisions	Negotiation
Hidden project costs	3	4	4	Margins Mark-ups Profit	Open book pricing Require transparency Reserve right to bid equipment

Legend: N/A not applicable; 1 low risk; 2 low-medium; 3 medium; 4 medium-high; 5 high

gy savings. The dollar value of energy savings can be adjusted to "present value" dollars. "Present value," or "discounted" dollars, are dollar amounts adjusted for the fact that, left invested over time, those dollars would generate income through interest earnings. This is sometimes referred to as the "time value of money." When making this adjustment the dollars are worth progressively less over time. The longer the payback; the greater the discount rate impact. The higher the interest rate; the greater the decline in dollar value. Future interest rates, therefore, become a risk factor that's hard to predict. The best source of assistance is the project financier, whose business relies in part on projecting interest and discount rates as accurately as possible.

The interest and discount rate risks can be limited by shorter contracts. Short contracts predicated on savings, however, force short paybacks. Short payback criteria remove the larger, longer payback, or "big ticket" items; e.g., boilers, chillers, window insulation, etc. from consideration. These big ticket items generally have greater savings persistence and, therefore, offer savings and environmental benefits for a longer period of time.

Energy Prices

Predicting the future price of energy is hard, if not impossible. The roller coaster we have been on since 1973 is apt to take some new dips and turns. To protect themselves, ESCOs typically guarantee the energy saved will cover the debt service *provided* energy prices do not go below a certain floor price. This provides some risk to the customer, but the obligation to pay the utility falls commensurably; so it is usually a wash.

The greatest risk at present, and probably the hardest of all to manage, in energy efficiency work is predicting how changes in supply costs will weigh against energy efficiency savings. For large organizations a fraction of a cent negotiated on the supply side of the meter can significantly outweigh any proffered efficiency economic gains. It is quite possible that negotiated or bid utility rates may fall lower than a stipulated floor price. This is most apt to happen in large industrial facilities, military bases and big commercial establishments where the owner's power to negotiate with the supplier is greatest.

Hidden Project Costs

When energy efficiency financing moves away from bid/spec, a major fear voiced by owners relates to hidden costs. Thus, the financial structure of a deal must be carefully examined for these costs.

On the other hand, it should be remembered that bid/spec is only as good as the quality of the specifications. Further, when "low bid" is required, the bidder is being asked to deliver MINIMALLY ACCEPTABLE equipment. Under traditional low bid, post contract savings, which usually accrue totally to the customer, can be very limited. Good business practices suggests the bidder will not invest in equipment, which would normally last longer than a specified and/or contract period, if he can help it.

It should also be remembered that the owner is seeking results; not just buying equipment. If all the owner wants is equipment, the organization is better off going directly to the vendor.

If management is concerned about hidden costs, it can always back out the numbers in proposed costs and identify the cost of the equipment and the cost of the services received. In addition to bid/spec equipment purchase options, there are mechanisms for establishing a framework for comparing prices with reference works, such as *Means**. If in-house expertise is not sufficient, outside consultation is a very good investment, especially on larger projects. If still in doubt, a customer can reserve the right to bid the equipment separately after reviewing the ESCO recommendations. An ESCO, however, must approve the specs and participate in final equipment selection if it is to offer a guarantee in such circumstances. A small organization, may not have this luxury if the savings opportunity is marginal.

An organization may also want to consider the open book procedures used in Canada, where prices are listed for categories of service as well as acquisitions.

In the final analysis, the net project financial benefit is a good indicator of the extent to which hidden costs exist *or really matter*. Too often potential customers get caught up in potential equipment costs and lose sight of the fact that the services and guaranteed savings are the desired result.

PROCEDURAL RISKS

Many of the procedural risks stem from insufficient attention to planning, selection procedures, inadequate equipment and maintenance

*Means, R.S., Means Mechanical Cost Data. Kingsten, MA

specifications as well as the lack of project supervision and staff training. Table 10-4 describes these risk factors and assigns the relative level of risk by performance contracting financing options. The major variables contributing to these factors and some mitigating strategies are also identified in the table's two right hand columns.

A review of Table 10-4 reaffirms that careful preparation and planning can effectively mitigate a significant portion of procedural risks. The variables and mitigating strategies are self-explanatory in the following table with four possible exceptions.

Project Management

The coordination between an ESCO and the customer is the prime responsibility of the ESCO's Project Manager and the organization's Energy Manager. To make it work, the responsibilities of the Project Manager should be clearly set forth in any ESCO proposal/contract as well as the qualifications of the assigned personnel to meet these responsibilities—in education, training and experience. The percentage of time the Project Manager will devote to the project each year of the project should be clearly stated. Assigning a new Project Manager, should a replacement be necessary, should always carry with it prior customer approval.

Facility Control Problems

Fear of loss of comfort is almost always the greatest concern among board members, management, and occupants. Contract stipulation as to the acceptable range of heating and cooling temperature parameters, relative humidity levels, air changes per hour, lighting parameters, and an override provision for special circumstances, etc., generally satisfy these concerns. The owner should set these parameters. It is well to remember, however, that the more control of the facility a customer requires; the greater the ESCO risk and, consequently, the greater the ESCO's share of the savings. However, both parties lose if the customer persists in maintaining an indoor environment that resists fulfilling the energy efficiency recommendations of the ESCO.

Quality Operations and Maintenance Training

The contract should clearly assign which parties will provide certain maintenance tasks. These tasks should be sufficiently detailed to establish the quality of maintenance to be expected. Check-off lists or a computer managed maintenance system can go a long way toward mitigating this

Table 10-4. Procedural Risk Framework

FINANCIAL RISKS FACTORS	PERFORMANCE CONTRACTING OPTIONS			MAJOR VARIABLES	MITIGATING STRATEGIES
	Vendor Financing	Shared Savings	Guaranteed Savings		
Facility selection procedures	3	3	2	Organization leadership	Organization admin. should establish
Poor equipment selection procedures	3	2	2	Planning, Specs - who writes - who approves	Quality in-house or consulting expertise
Audit is not sufficiently comprehensive	N/A	4	2	ESCO selection Auditor selection	Quality in-house or consulting expertise Length of contract
Project management inadequate	N/A	2	2	ESCO selection Project Manager selection	Careful planning & selection
Facility control problems	2	2	2	Contract provisions	Specify acceptable parameter s in contract
Quality of maintenance; training	2	2	1	Specification for maintenance Trainer's abilities; cost	ESCO selection Project Manager
Emergency response provision parameters in contract	1	2	2	Contract provisions	Specify acceptable
Termination conditions & values	3	3	2	Contract provisions	Negotiations Legal abilities
Schedule adherence	1	1	1	Delays	Check past practice of vendor, ESCO penalties, bonuses
Intrusion/interruption	1	2	3	Variable contract language Size of project	Mitigate through careful attention to operational needs Project Manager

Legend: N/A not applicable; 1 low risk; 2 low-medium; 3 medium; 4 medium-high; 5 high

risk for both parties.

Training for O&M staff should be offered by ESCOs on installed equipment at a minimum, and on related energy consuming equipment wherever feasible. In most circumstances, both parties gain if the training is offered annually for the life of the contract.

Emergency Response Provisions

The level of risk associated with emergency response will vary considerably based on the measures installed and how critical they are to the operation. The response time on a burned out lamp, for example, is usually not as critical as when a chiller goes down on a hot summer day. The required response time necessitates an assessment of how critical the equipment is and how long the facility can coast. Typically, ESCOs will rely on local distributors as their first line of defense.

Keeping Risks in Perspective

Whenever this much attention is paid to potential customer risks, there is the inherent problem of enlarging customer perception of the risks to be incurred. And consequently stirring fear in the hearts of the less venturesome. It is appropriate, therefore, to come full circle and once again state that performance contracting is *a risk shedding opportunity*. The owner should incur less risks in performance contracting than in any other energy efficiency financing approach. The biggest risk for a owner, in fact, is selecting the right ESCO.

The Big Risk: Getting The Right ESCO

The risks associated with selecting an ESCO are about the same as in other energy efficiency financing approaches: insufficient attention to planning and selection procedures at the outset. Careful preparation and planning can effectively mitigate many risks.

Planning and preparation should include:

(1) deciding on the desired results and determining the criteria, which will reveal the firm best prepared to deliver those results;

(2) developing effective selection procedures, including specifications and / or request for proposal language, and evaluation procedures;

(3) establishing contract language, which reasonably protects the organization;

(4) negotiating reasonable contract terms; and

(5) realistically examining capabilities of staff and retaining quality outside consultation whenever needed. [Note: In a project of sufficient size and value, these consultation costs can be assigned to the project.]

Since ESCO selection is such a major risk, the reader may wish to refer back to Chapter 4, " Partner Selection: From Both Sides of the Fence."

In the final analysis, risk management answers two questions: What do we get out of it if we proceed? If we go ahead, how can we control any associated problems and costs? When energy efficiency is considered, the answers for an organization may vary; but the principles are the same.

Chapter 11

ESCO Risks and Management Strategies

When you peel away the trimmings, the heart of performance contracting is risk management. The ESCO, which can most effectively manage those risks, has a tremendous advantage in the market place.

By definition, an ESCO offers guarantees. Risks are inherent to the guarantee process. How those risks are identified, assessed, managed and mitigated is absolutely critical to an ESCO's success.

To grow and prosper, an ESCO needs a very pragmatic approach to risk management. Two avenues are needed to meet this goal: a) identification of the most common procedural/administrative risks ESCOs face and techniques for minimizing their impact on project profitability, and b) determination of measure-specific risks and ways to cost effectively mitigate their impact on projected savings.

In the final analysis, the ability of the ESCO to identify risks, assess their implications, determine appropriate mitigating strategies, and manage them effectively makes a decisive difference in the level of services and savings an ESCO can offer.

Performance contracting works best when the savings revenue stream funds the project. The more secure that revenue stream; the more an ESCO can do in a given project. The ESCO, who effectively manages project risks, can deliver a better program to the customer *for the same level of investment*. The customer benefits as more is invested in equipment and services, and more savings are achieved. Consequently, the ESCO with a secure revenue stream earns a greater profit and attains a solid reputation for delivering on its guarantees.

ESCO Risk Vulnerability

The greatest points of risk vulnerability have emerged as the ESCO industry has matured. Today, the experienced ESCOs know with more precision which performance contracting aspects need careful scrutiny and management. A staggering number of horror stories have helped to underscore critical risk aspects. How these aspects manifest themselves, their magnitude and mitigating strategies may vary by market or by country culture, but there is an underlying consistency. The major risk factors that must be managed by an effective ESCO have been established: customer pre-qualification; project development; energy audit quality; equipment selection and installation procedures; commissioning; operations and maintenance practices and training; measurement and savings verification; and project management over the life of the project.

CUSTOMER PRE-QUALIFICATION

For the firm just entering the ESCO business, the most obvious customer qualification seems to be energy savings potential. But that is far too simplistic—even dangerous. The best potential savings opportunity in the world does not help if the customer goes out of business and is not around to make payments to the ESCO.

Since performance contracting is primarily a financial transaction, the financial condition of the customer can exceed savings potential in importance. The customer must remain in business for the life of the project if the project is to remain an economically viable endeavor. For the ESCO to stay in business, its fees must be paid. To the extent that the customer incurs debt to purchase the equipment, that customer must be in a position to meet the debt service obligations.

Of nearly equal importance is the impact the customer's management and staff can have on the project. As noted throughout the book, the "people factor" can make or break a project that otherwise seems very attractive. Ways to defeat our best energy efficiency efforts abound.

Qualification Criteria

To manage the customer risk effectively, each ESCO needs to determine the criteria a customer must meet to be an attractive candidate for performance contracting. The ESCO must also develop the questions, and perhaps a survey form, to ascertain if the criteria are met. The final and most difficult piece of the management strategy is to know *when to walk away*. It is hard for the sales and marketing people to turn their backs on a potential customer that just oozes energy savings potential; but if key criteria are not met, the project won't work.

Pre-qualification criteria can usually be divided into three categories: financial/economic factors, facility/technical factors and people factors. These categories may differ in matters of emphasis with the customer, project and/or measures. For example, people criteria are not very important if the only measure being implemented is roof insulation. Specific criteria may need to be reevaluated and refined with changing conditions or different business models.

Financial/economic factors:

- creditworthiness,
- organization's longevity, stability,
- organization's business prospects, and

- supporting documentation.

Specific markets and given customers will prompt additional financial criteria. Many financial institutions now have stipulated conditions, even forms, to guide ESCO sales people.

Facility/technical factors:
- building age and useful life,
- building function, potential changes,
- occupancy schedules, density
- condition of mechanical equipment,
- existing level of maintenance,
- annual utility bill, and potential project size, and
- energy and cost saving potential.

These criteria will be addressed in depth during the investment grade audit (discussed below) but a preliminary judgment must be made as to whether these criteria are met sufficiently to warrant further project development. The preliminary judgment is usually made with a "walk-through" scoping audit.

People factors:
- top management commitment,
- multi-level involvement, team approach,
- ability to consistently make decisions in a timely manner,
- evidence that the concept and benefits are understood,
- management has needs, wants, wishes that the project will serve, and
- O&M manpower, abilities, training needs, attitudes.

Assessing these criteria is very difficult, auditors do not always have the needed analysis skills and portions of this assessment are quite subjective. However, ESCOs should do all they can to quantify people factors from maintenance to top management.

The level of top management's commitment is very important, but an absolutely critical factor is the attitude of the O&M staff. If the director of facilities or maintenance announces that a performance contracting project will not work in "his" building, the ESCO can be assured *it will not work*! The person, who has responsibility for the switches and valves, can

defeat the best designed project in the world.

Pre-qualification of customers is both a science and an art. The ESCO that does not master this area will spend a lot of time and money chasing empty promises. Time and money that could be spent seeking viable customers. Salespeople also need guidance as to the timing of securing answers to certain criteria as well; so they know when to cut their losses. For example, if the financials are not clear and adequately documented by a specific stage in the sales cycle, internal counsel should be sought before any more sales time is expended.

PROJECT DEVELOPMENT

From the first call through the Project Development phase, the ESCO is investing money in a potential project. The biggest enemy in the ESCO business is time. The more protracted the Customer Qualification and Project Development phases are; the more costly.

The Project Development phases, therefore, should be carefully planned and smoothly executed. Each risk associated with this phase should be acknowledged and addressed. A major concern is getting the "buy" decision. This means that a very early determination regarding customer pre-qualification, the ability of key decision makers in an organization to decide and act, may determine if further work is warranted.

"BUY" Signals

A crucial early step is to determine who are the decision-makers and the influencers in making a buy decision. Influencers, such as energy managers and directors of maintenance, can often slow down or kill a deal, but they seldom sign the contract. The key considerations at this stage are: (1) who are the decision-makers; and (2) can the management make a decision in a reasonably short time. A protracted sales cycle, or a never completed deal, not only delays the day when revenues can flow, but ties up personnel that could be used to make another project happen. Identifying buy/no buy signals is a critical part of an effective marketing strategy. Unfortunately, the ability to read such signals usually comes with experience—sometimes very painful experience.

By the time the Project Development phase is reached, more ESCO funds have been expended; but no revenue has been realized. Compressing the sales cycle through successful sales strategies is essential.

One ESCO has successfully used a Decision Schedule that lists; (a) the tasks leading to a contract, (b) the responsible party for each task, and (c) the task completion dates. After the concept sell is accepted, both parties develop and sign the Decision Schedule. Then, should one party lag, a courteous letter saying, "We cannot meet our date for... Task 7... completion because your financial officer still has the material..." generally gets the process moving again. The Decision Schedule should be signed by both parties, but should not be legally binding.

Deal stoppers during the Project Development phase include:

- Finding the project cannot be financed;
- Recognizing the project cannot be paid for out of the projected savings;
- Determining projected savings cannot be measured effectively; or
- Realizing the savings baseline is too dynamic to be managed.

Anticipating these issues and taking steps to resolve them may avoid such problems. Or, at least avoid the money hemorrhaging on a futile effort. Pre-approved project financing, for example, will make it clear early in the process exactly what the financier expects.

An effective way to move from the Project Development phase to the Investment Grade Audit (IGA) is to prepare a concept report with a benchmark energy analysis and rough numbers as well as the agreed upon project objectives. This document then serves as the basis for a Planning Agreement, which is a preliminary agreement requiring the owner to pay for the IGA if the project does not go forward.

INVESTMENT GRADE AUDITS

ESCOs have learned, often through painful experience, that the traditional energy audit is just not good enough. An audit that incorporates all the name plate data, run times, etc. is only the first step. If an ESCO is going to invest its time and money in developing a project and laying its guarantees on the line, then it must have quality information to justify the investment—and the gamble.

The traditional energy audit has been a "snap shot" approach that typically assumes all conditions on audit day will remain static for the projected payback time of the measures. But buildings are seldom—if ever—static. They are typically dynamic places with changing functions. What's more, they're populated by people who simply will not behave in predictable, consistent ways.

For investors and energy service companies that rely on energy and operational savings to ultimately fund retrofits, the 1970's energy audit has increasingly fallen short of the mark.

We have historically skirted the implications of the human element in energy auditing. "Paybacks" have been assigned to certain measures in multiple applications when we knew full well they would not perform in exactly the same manner under differing conditions. Remember, as stated earlier, that up to 80 percent of the savings in an effective energy management program can be attributed to the energy efficient practices of the O&M personnel. In other words, as little as 20 percent of the savings could be attributed to the actual hardware, but we have continued to make calculations as though a piece of hardware was going to always operate in the same fashion even under vastly differing conditions.

Paybacks based solely on the cost of acquisition and installation are myths. Very misleading myths.

Frequently, the impact of existing energy-related equipment on newly installed equipment has also been neglected. Back when we were all on a steep learning curve, a controls company attempted to duty cycle a 100 year old boiler in Providence, Rhode Island. With 20/20 hindsight, we now know that was not a good idea.

Over the years, experience has taught those in the performance contracting industry that guarantees require more precise calculations of the conditions which surround newly installed energy efficient equipment, and the unpredictable element people bring to the equation.

Today, those who wish to predict savings with any degree of confidence must turn to an ***investment grade audit***.(IGA). The following paragraphs offer a brief summary of the IGA, its characteristics and benefits. The reader, who wishes more information, should read *Investment Grade Energy Audits: Making Smart Energy Decisions* by Shirley J. Hansen and James W. Brown, available from The Fairmont Press.

An IGA incorporates the name plate data, run hours, and other information that goes into a traditional audit. Then, a *risk assessment component* is applied which assesses conditions in a given facility, and more importantly, looks at the human aspect. The challenge is to determine how the proposed measures will really behave *over time* given the probable future conditions.

The human factor must not only be assessed, but paired with potential energy measures to ascertain the impact occupants, management, maintenance and operational behavior will have on the energy efficiency measures. For example, measures, which are practically people impervious such as insulation, can be looked on more favorably, especially in facilities where the human factor receives a relatively low score. On the other hand, measures such as controls—particularly if overrides are readily accessible—carry a greater risk because of the human factor. All of these factors must be considered while the payback and predicted savings are being forecast.

An IGA is far more demanding, requires greater skills, and necessitates some subjective judgment. The auditor must weigh many key factors, including:

- management leadership and its commitment to energy efficiency;
- the resultant occupant behavior based largely on management's visible commitment;

- the manpower, skill and training needs of operators and maintenance staff;
- the level of equipment sophistication the O&M staff can operate effectively;
- the condition of energy-related mechanical equipment;
- repairs and replacement budget provisions; and
- the attitude of O&M personnel towards the energy program.

Once these and other human factors are weighed, an IGA requires that they be converted to risks with price tags. The whole financial structure of an energy project, especially those with savings guarantees, must allow for these risks and the cost of appropriate mitigation strategies.

Woven through all the technical/human consideration, is the money component. Life cycle costing, complete with net present value calculations, must be part of an IGA.

General facility upgrades and needed equipment replacement often drive projects. Energy efficiency benefits from these changes are becoming a carefully calculated part of the investment package. The measures must yield the calculated benefits if the package is to be economically viable. An assumption that things will stay the same just doesn't cut it anymore.

The term "investment grade" was coined to also reflect the fact that an audit report should provide the owner with an investment guide. Facilities and processes are major portions of any organization's investment portfolio. An IGA report should clearly document how the recommended measures (investments) will enhance that portfolio's value.

Engineers who can perform a quality IGA bring in the money. They are in short supply and in increasing demand. The search is always on for auditors, who can perform a quality IGA, thereby reducing the ESCOs' risks. In fact, many ESCOs now charge owners a premium for an investment grade audit if the project does not go forward. This charge is incorporated into a preliminary Planning Agreement as discussed in Chapter 7 "Contracts."

When all is said and done, a quality energy audit must stand up to the careful scrutiny of bankers and other investors. An IGA is at the heart of a "bankable project." Hence the term, *investment grade* audit. Any thing less no longer adequately serves the owner, the contractor, or the investor.

A word of caution: "Investment Grade Audit" has such a nice ring to it that many engineering firms and ESCOs have changed their audit name but not their auditing procedures. One firm has even claimed to have been doing IGAs since the 1970s. Since auditing was in a very embryonic stage at the time and the term had not been coined, one must doubt the claim.

To test such claims, owners should ask firms for their track record on predicting savings. Their predictive consistency (predicted savings vs. actual savings achieved) should be between 90-110%.

THE CRITICAL BASEYEAR/BASELINE

Inadequate baseyear documentation can become a bubbling caldron that only gets hotter through the project years. A clearly documented baseyear, as part of the IGA and the contract, deserves special mention, for it is a key part of effective project management and savings verification.

An auditor cannot accurately predict savings without knowing exactly what the existing conditions are. Too often, however, those conditions are not clearly set forth and signed off by the owner. Then, the squabbles start over how long the lights used to be on… and they build from there.

In addition to the energy consumption, all pertinent operating conditions that contributed to that consumption need to be ACCURATELY noted. Stressing "accurately" seems rather redundant, but we still find auditors who ask management how long the lights are on… and believe what they hear! Or, they ask the custodian and take the response as gospel. Nine times out of ten, the numbers from management and the custodian don't even agree. This is one time when a "little black box" really works. The portable data logger, if used judiciously prior to the audit calculations, can yield valuable data and avoid later misunderstandings. The same care in documenting other operating conditions is needed.

EQUIPMENT SELECTION AND INSTALLATION

Equipment selection should grow out of the audit recommendations, as it would in any energy efficiency project. The latitude given the ESCO by the owner in this selection process significantly reduces the level of risk the ESCO carries. The greater the customer's say in the matter; the greater the ESCO's risk is apt to be.

At the very least, the ESCO needs to be fully involved in preparing the equipment specifications. If the owner insists that the equipment be bid to assure competitive prices and/or cost transparency, the ESCO should participate in the final selection process. In this instance, one party may narrow the field of acceptable bidders to two or three; then the other party makes the final selection.

In much the same fashion, the ESCO must have some control over the subcontractors, who will install the equipment, to assure that it is installed correctly and will operate near design.

Risks associated with equipment selection and installation often put the owner and ESCO at odds. The ESCO would prefer to function as a general contractor and have full responsibility for all aspects of project implementation. The owner is rightly concerned about the quality of equipment installed on his premises. Frequently, the use of local subcontractors is also important to the owner. In part, because a particular subcontractor is already familiar with the facility and has the owner's confidence. Politically, it is often attractive to use local labor for community relations or to bolster the local economy. Working out an amicable owner/ESCO arrangement at this point can help set the stage for a true partnership—and a successful project.

The construction manager can play a key role in assuring the right equipment arrives at the right time and is installed by those qualified to do so. Chapter 14, "Productive ESCO/Vendor Relations," offers ways to involve the vendor, which in turn will prompt more timely service. For the owner and ESCO, a construction manager working closely with the owner's lead person, can effectively hold down risks and costs. An effective construction manager oversees the process, minimizes the delays and the associated costs, and keeps the whole process on schedule. Critical to a retrofit project, the construction manager also keeps intrusion in the organization's procedures to a minimum and keeps existing building and industrial processes in operation as much as possible.

There is some latitude for negotiations between the customer and the ESCO regarding equipment and installer selection, but not much. The customer should keep in mind that the more control the organization keeps in this process; the greater the ESCO risks. As we will see later in this chapter, the greater the ESCO risks are; the less the project benefits the customer. In the final analysis, as long as an ESCO guarantee is involved, the owner will pay to some degree for any measure of control the organization chooses to keep.

COMMISSIONING

Difficulties with existing buildings and increasing concerns about indoor air quality have underscored an owner's need for commissioning. Gradually, ESCOs have also recognized that commissioning can help reenforce the partnership relationship at the outset, and assure all parties that the design criteria have been met and all installed equipment is operating correctly. Since commissioning includes a performance verification procedure, it can also serve as a basic procedure for the measurement and savings verification process.

The concept of commissioning has been borrowed from the shipping industry. The idea grew from recognition that new ships operate at sea and must operate as designed. Service calls and system downtime are not attractive options when this floating "building" is a long way from port.

While land-locked servicing and downtime problems may not be quite so dramatic, occupant health, safety, comfort and productivity has encouraged commissioning as a critical part of owner acceptance of newly installed equipment or new construction.

A well-designed commissioning process can assure that:

- The new equipment has been tested and its performance verified in the presence of owner's staff and the ESCO's Project Manager; so those responsible for its future operations and savings potential concur on its acceptable performance;

- Calibration procedures, fine tuning, and routine maintenance are performed and clearly understood. Integral to this process is the scheduling of services and the identification of responsible staff and contractors who will perform the necessary procedures;

- The intended benefits and the optimum operating procedures are understood by all involved, including management, O&M staff and occupants. Without this clear understanding, faulty operations may go undetected; or difficulties may be ascribed to the new equipment which are unwarranted;

- All documents; e.g., manuals, cut sheets, building drawings, specifications, etc., should be conveyed to those who will operate and maintain the energy consuming systems; and

- Another cornerstone is laid in the growing sense of partnership between the owner and the ESCO.

For some, commissioning may seem to be inconsistent with the "turnkey mentality" of the building industry. It is, however, entirely compatible with a performance contract partnership where maximum comfort and savings are the goal.

In recent years, the commissioning function has been separated into retrocommissioning, continuous commissioning and recommissioning. The staff at Portland Energy Conservation, Inc. and the Oak Ridge National Laboratory issued a report entitled, "A Practical Guide for Commissioning Existing Buildings" in which they defined these four difference commissioning types or levels:

"*Commissioning* is defined in ASHRAE Guideline 1-1996 as the process of ensuring that systems are designed, installed, functionally tested, and capable of being operated and maintained to perform in conformity with the design intent... commissioning begins with planning and includes design, construction, startup, acceptance, and training and can be applied throughout the life of the building." The term commissioning, used in this sense, is generally accepted as an analysis of new construction projects.

Retrocommissioning (or existing building commissioning) is defined as "an event in the life of a building that applies a systematic investigation process for improving and optimizing a building's O&M." The report goes on to say that "its focus is usually on energy-using equipment, such as mechanical, lighting, and related controls" and "is applied to buildings that have *not* previously been commissioned." Finally, the report states that "The recommissioning process most often focuses on the dynamic energy-using systems with the goal of reducing energy waste, obtaining energy cost savings for the owner, and identifying and fixing existing problems."

Continuous Commissioning is defined as a process much like Retrocommissioning with objectives that are essentially the same. The difference, however, is that Continuous Commissioning "more rigorously addresses the issue of persistence. A key goal is to ensure that the building systems remain optimized continuously."

Recommissioning "can occur only if a building was commissioned at some point in its life... recommissioning is a *periodic event* and *reapplies* the original commissionings tests in order to keep the building operating to design or current operating needs."

Operations and Maintenance Practices

The "up to 80 percent" figure attributable to operations and maintenance (O&M) energy efficiency practices referenced earlier in this chapter is also very scary to ESCOs, who are guaranteeing the energy saving performance of the equipment. To help manage this risk, many ESCOs prefer to maintain the equipment themselves; or, designate a contractor, who is also under some performance guarantees, to do it for them.

Setting the O&M tasks and schedule by measure is a key management function whether the ESCO's own personnel perform the work, it is assigned to a contractor, or the customer's staff performs the needed tasks. In every case, some policing of the facility or process is essential to assuring the tasks have been performed effectively and on schedule. The risks are greater and the policing more imperative when the O&M functions are further removed from the ESCO.

Some owners think they will save money by having their own personnel perform the O&M tasks. This is questionable and comparative calculations should be made. In such comparisons, consideration must be given to the size of the risk cushion the ESCO must set aside to cover the greater risks associated with O&M work provided by the owner.

While making such calculations, customers should be aware that ESCOs frequently make as much or more from the maintenance contract than they do from the savings and related services. If the maintenance services revenue is denied an ESCO, the project will be less attractive. Some of the lost revenues may be made up in other aspects of the project. Or, owner maintenance may add sufficient risk for the ESCO that the risks exceed the benefits and the project is no longer viable.

Some of the maintenance risks, especially those performed by a subcontractor or by an owner, can be mitigated by using a computer-based maintenance management system (CMMS). These programs are flexible enough to be tailored to the exact measures implemented and have hundreds of tasks detailed for various energy savings measures. It is an automated way to police the quality of the maintenance as well as attribute and document the inadequacies that cause measures to fall short of their promised savings. Both the ESCO and the owner are urged to consider CMMS when a broad range of equipment has been installed and a myriad of tasks are necessary. Software that already contains such a range of equipment and the associated tasks should be used. Research reveals that if these manufacturer's data entry tasks are left to the purchaser, the success rate for CMMS use goes down to a dismal 20 percent.

MEASUREMENT AND VERIFICATION

A well-planned measurement and verification protocol (M&V) can offer strong risk mitigating opportunities. Whenever money changes hands based on savings, some validation of the savings is necessary.

Chapter 5, "An ESCO's Guide to Measurement and Verification," describes in detail M&V's current status and recommended protocols. It should, however, be stressed here that inappropriate M&V can actually add to performance contractor's risks. When M&V plans are not appropriate to the measure, yield information that is questionable, or are too costly for the value of the data provided, they work against a sound performance contract. And escalate the risks.

Conversely, a good M&V protocol, agreed to in advance by both parties, not only serves as a valuable tool but offers a risk mitigating opportunity.

PROJECT IMPLEMENTATION

All too often, new ESCOs talk about a project being completed once the equipment has been installed and accepted. That's from the old direct purchase/construction project mentality. The successful ESCO and its partner realize the project is *only beginning* when the "stuff is in the ground."

Due to the importance of effective project management through all the years of the contract, Chapter 8, "Where the Savings Are: Project Management," has been devoted to this key aspect. The point in bringing it up in this context is to stress the risk implications inherent in this phase of the project.

The principle difference between a partnership with open and effective communications between ESCO and customers, and one headed for the courts, is a quality Project Manager. This is followed closely by the qualities of the customer's Energy Manager—and how well the two project leaders work together.

The ESCO's interest in achieving the full project implementation benefits is largely influenced by where the ESCO's fee is placed in the financial structure. The mechanics of this are discussed later in the chapter. There is a bottom line, however, in the risk area, which potential customers should be fully aware: *The manner in which an ESCO takes its fee out of the project influences its interest in full project implementation and achieving excess savings over the life of the project.*

At least one ESCO buries its fee in the guarantee package and claims not to take a share of the savings. This procedure certainly reduces this ESCO's project risks significantly. Such an arrangement, however, offers no incentive for the ESCO to work with its partner beyond achieving the guarantee. As long as the savings don't drop below the guarantee level, this ESCO carries little concern about how the project will perform over time. There is a price, however, as the ESCO incurs a greater risk of customer dissatisfaction.

The owner is better served by an ESCO, which takes some or all of its fee out of the excess savings over and above the guarantee.

Overall the ESCO is best served by taking enough of its fee out of the initial financing to cover the costs it has already incurred. In such situations, the ESCO's basic revenue needs are satisfied and the portion of the fee coming from the excess savings offers the ESCO an incentive to assure the excess savings happen.

MANAGING RISKS THROUGH THE FINANCIAL STRUCTURE

Risk management strategies impact the bottom line—for all parties. The way ESCOs use the financial structure to manage risks, described below, substantiates the assertion that performance contracting is primarily a financial transaction.

ESCOs do not guarantee all the savings they expect to achieve in a building and/or industrial process. The level of guarantee is influenced by the perceived risks.

Through the audit process, the ESCO determines what a reasonable level of potential savings will be. Then a percentage of that "reasonable level" is guaranteed as shown in Figure 11-1. The guarantee covers the normal construction costs; e.g., design, acquisition, installation and interest on the money. The risk cushion established by the ESCO (perhaps unknown to the customer) will typically vary the guarantee level from 50 to 80 percent of the expected savings potential. Some ESCOs insist their sales engineers not exceed an 80 percent guarantee mark. A few ESCOs will go above the 80 percent mark based upon very secure measures, exceptional confidence in the auditor's capabilities, or with a highly qualified customer. The difference between the potential guarantee and the actual guarantee is the "risk cushion." The "risk cushion" is the ESCO's protection against the real and perceived risks. If a project performs as expected, there will be excess savings above the guarantee. This amount, shown in Figure 11-1 as "excess savings" is sometimes referred to as "positive cash flow."

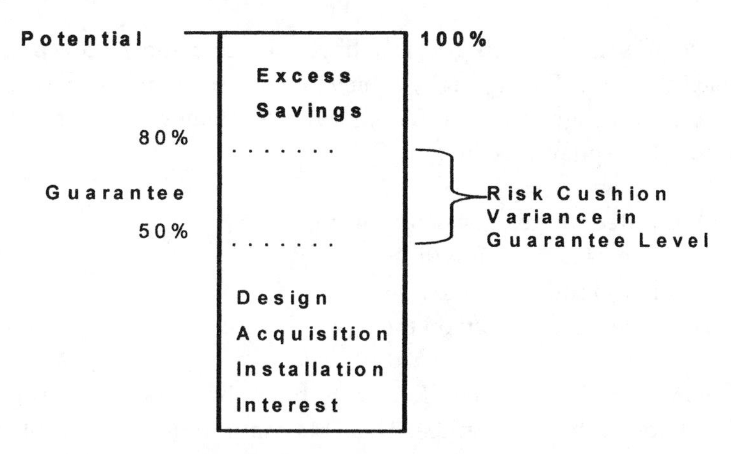

Figure 11-1. Managing Risk through the Financial Structure

The level of the guarantee will vary both by the perceived customer-associated risks and by the ESCO's confidence in its own ability to assess the risks and develop effective mitigating strategies. As depicted in Figure 11-2, two companies may assess the same facility, calculate the same savings potential, but perceive the risks quite differently.

The difference between Company A's and Company B's guarantee level, as shown in Figure 11-2, is their respective risk analysis capabilities. Whether it is capability or caution, Company B will deliver a smaller project with the same level of customer investment.

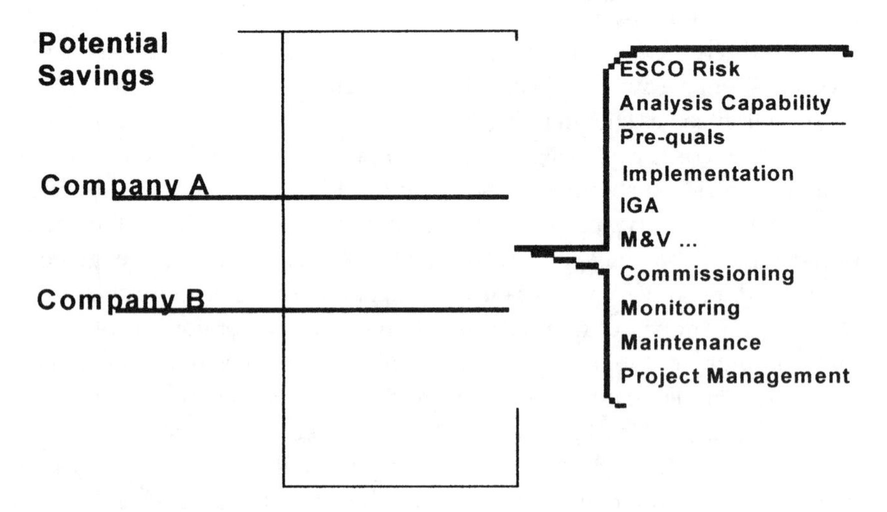

Figure 11-2. Risk Strategies and the High Price of Caution

The IGA risk analysis goal is to have predicted and achieved savings consistently close. Surprisingly, achieved savings should exceed projections by only a slight amount. ESCOs, who consistently show savings far in excess of the guarantee are:

a) limited in their risk analysis capability;
b) have basic auditing problems; or
c) are too cautious, and
d) are leaving money on the table.

Some ESCOs have been known to brag about the 40 to 50 percent achieved above their guarantee. This is nothing to brag about. Money is being left on the table. Such "achievements" send a clear signal of an over-cautious ESCO, which is hurting itself and its customers.

An example will demonstrate the losses that come from cautious ESCOs, who are so pleased with their excessive positive cash flows. Referring back to Figure 11-2, let's assume a customer's energy bill is $10 million per year and Companies A & B both identify energy efficiency measures to reduce the savings by 25 percent, or $2.5 million per year. Company A decides it can place its guarantee level at 75 percent of the expected savings ($2.5 million X 75% = $1.875 million) while Company B puts its guarantee at 60 percent ($2.5 million x 60% = $1.5 million). The level of guarantee per year from the two companies will differ by $.375 million as shown in Figure 11-3. If the payback period is four years, the level of investment in equipment can be $1.5 million greater for Company A... and for the owner. If Company A is right and the mitigating strategies and greater investment in equipment yields $.375 million more per year, the guaranteed savings over 10 years will be greater than Company B's guarantee by $3.7 million. The total guaranteed benefit to the owner would be $5.2 million greater with Company A. Even if both companies, achieve the same potential level of savings the owner will have gained $1.5 million more in capital investments with Company A. And ESCO "A" will have sold and implemented a larger project.

Finally, it should be noted that the implementation of a bigger package by Company A enables that company to be more assured of achieving the predicted savings and even exceeding those predictions.

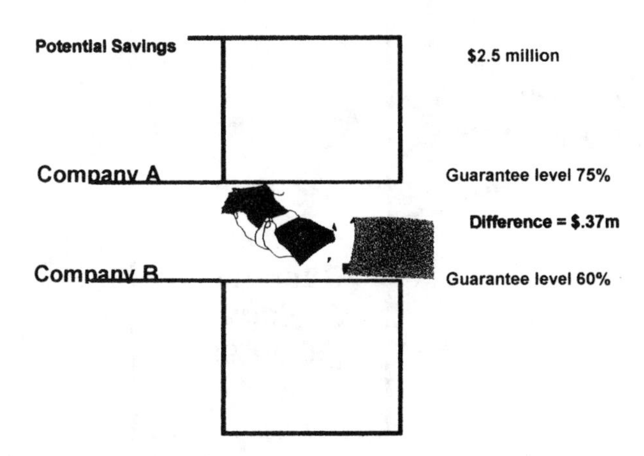

Figure 11-3. Effective Risk Strategies; Implications to Owner

The ESCO Fee

All ESCOs charge a fee for their services. Some try to suggest otherwise by urging customers not go with any ESCO that wants a share of the savings. The implication is that this ESCO does not take any money from the savings for its services. Not true. A performance contracting project is funded from the savings. No ESCO, to our knowledge, is totally altruistic. Certainly not the one making this claim.

If we examine three scenarios, the significance as to the ESCO fee placement to the owner and the project become very evident. The "altruistic" ESCO #1, as noted above, takes all of its fee out of the guarantee package. ESCO #2 takes some fee from the guarantee package and some from the positive cash flow. ESCO #3 takes its fee entirely from the cash flow. The implications for the ESCO and the owner in these three scenarios are significant.

ESCO #1

The ESCO #1 gets all of its money from the financier up front. The ESCO benefits from the fact that the money need not be discounted (See

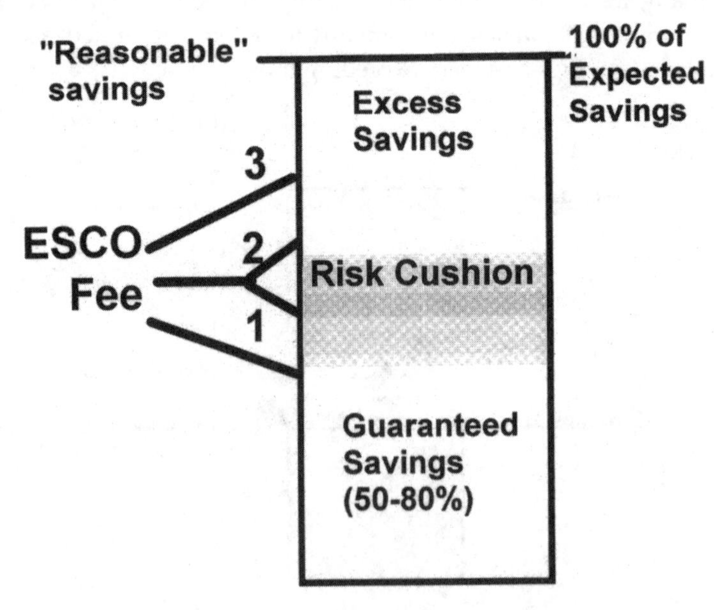

Figure 11-4. ESCO Fee Placement

Chapter 6 for impact of discounting). The return on investment is more secure and the risks are reduced. As long as the guarantee is met, ESCO #1 has little concern as to how the project performs. Customer dissatisfaction is typically higher with this approach.

As discussed in the project implementation section, ESCO #1's assistance to the owner to achieve any savings, above the guarantee, is less certain. Since the ESCO fee for all of the years of the contract are buried in the financing, the owner pays interest on the ESCO fees for multiple years. If, for example, the ESCO fee is $100 thousand per year on a 10 year contract, the owner is paying interest compounded annually on $1,000,000 that it would otherwise not have to pay.

When the ESCO's sales pitch states the owner gets "all the savings," or "all the excess savings," or "all the positive cash flow," customers should recognize they may get more immediate savings but they are apt to pay for it in interest charges and lost capital investment.

ESCO #2

This ESCO places some of its fee in the guarantee package and gets some of the fee from the excess savings. At the time of the construction funds are paid, the ESCO recovers all sunk costs and margins and/or profits on those expenditures. For the ESCO, the portion in the guarantee package is a secure return on investment and need not be discounted. The risks are somewhat reduced. The ESCO has some interest in assuring the customer achieves some positive cash flow. Customer satisfaction is apt to be better than with ESCO #1.

For the owner, ESCO #2 is more apt to be an active partner than ESCO #1 for the life of the project. The portion of the ESCO fee in the financed package will be smaller; therefore, the interest costs will be lower. The owner will receive most, but not all the excess savings. The savings, however, due to continued ESCO involvement are apt to be greater than with ESCO #1. When an ESCO proposes to take a relatively small portion of the excess savings, the type #2 ESCO is usually involved.

ESCO #3

This ESCO takes all of it fee out of the excess savings. The risks for the ESCO are higher if excess savings are not achieved. The fees are paid out over the life of the contract and need to be discounted. The ESCO is a very active partner for the life of the project.

The owner pays no financing charges or interest on the ESCO fee.

The ESCO, however, typically takes 100 percent of the excess savings to a specified dollar amount; so the owner does not have the potential for immediate positive cash flow. The active partnership of the ESCO generally yields greater excess savings.

There is usually greater cost transparency, as a type #3 ESCO clearly spells out exactly how its fee will be paid from the excess savings.

Hybrids of these three scenarios may also be offered. Potential performance contracting customers can rest assured that somewhere in the financial structure is an accommodation for the ESCO fee. Every ESCO gets a fee, even those who characterize themselves as non-profits.

In the final analysis, how risks are assessed and managed financially affect all parties. It is in the interests of the customer, the ESCO, and the financier to work together to develop an effective risk management strategy.

The ESCO, which identifies and manages its risks effectively, will deliver better projects, be more competitive and show strong company growth. In short, ESCOs that effectively manage risks are more successful.

Chapter 12

Energy Efficiency and Indoor Air Quality

In the 1970s, with lines at the gas pump and "energy crisis" on everyone's lips, it became common practice for facility people to cover outside air intakes in the hopes of reducing energy consumption and cost. Such actions, however, often had a negative impact on the quality of the indoor air. Unfortunately, out of such misplaced energy conservation efforts came an assumption that we couldn't have both energy efficiency and indoor air quality.

Today, we know that indoor air quality (IAQ) risks associated with energy efficiency* are more perceived than real. But the persistent fears that energy efficiency measures may have a negative impact on IAQ have affected owners, energy efficiency vendors, engineers, consultants, and energy service companies. These fears have increased perceived risks, created sales resistance and changed the financial dynamics of many projects. Recognizing that such fears do exist and treatment is required is a critical step in achieving both energy efficiency and indoor air quality.

Energy efficiency advocates need to know where those fears originated and how they manifested themselves. An even greater need is an assessment of the real relationship between IAQ and energy efficiency (EE).†

Over the past two decades, nearly every IAQ magazine article in the second or third paragraph has mentioned the energy crisis of the 1970s, the resulting tight buildings, and the range of IAQ woes. Readers of these articles have been left with the persistent impression that, as energy pric-

---

*In considering indoor air quality, the distinction between energy efficiency and energy conservation becomes critical. While energy conservation, by definition, means to use less, energy efficiency is defined as using the needed energy as efficiently as possible.

†More information about the relationship between IAQ and energy efficiency can be found in the book, *Managing Indoor Air Quality*, by H.E. Burroughs and Shirley J. Hansen.

es soared in the 1970s, those "unthinking" owners and facility managers tightened buildings to save dollars and left the poor occupants sealed in boxes gagging on stale air filled with pollutants. The popular press informed us repeatedly over the years that we can't have both EE and IAQ. THEY WERE WRONG!

ESCOs have found market resistance based on this fallacy and have had to assume greater risks in making comfort or health assurances.

Facility people have been confronted with a dilemma. On one hand, poor indoor air quality can hurt productivity, lose tenants, and even create health/legal problems. On the other, there is the constant demand to run facilities as cost-effectively as possible, which places energy efficiency as a high priority. Guidance is needed.

In the Beginning...

Our first question, then, is: Where did the idea originate that the energy efficient building was at fault? Somehow the idea that a tight building is not good and uses only re-circulated air has permeated indoor environment thought processes. Ventilation has become THE answer. For over ten years, the ventilation disciples have almost convinced us that's the way to go. Open the window! Air will just "naturally" get better.

But has it? Will it? The answer, unfortunately, is: **Not necessarily!!**

FINDING ANSWERS

ALL THAT "FRESH" AIR

When increased outside air is proposed, the ESCO must consider the quality of the air outside. What if, for example, the air outside is worse? The fresh, *natural* air sounds so wholesome. Try stepping outside of the United terminal at O'Hare International Airport; the air outside is much worse than the air inside. There is no "fresh" outside air for the O'Hare facility people to bring into the terminal. Natural ventilation would be a disaster.

Opening the window is frequently not an option for owners—or ESCOs.

If we ask hay fever suffers about opening the windows and letting in all that wonderful fresh air, between sniffles they will argue strongly against it.

VENTILATION AS THE ANSWER

Ventilation is not always the answer. To "clear the air" about the relationship between IAQ and energy efficiency, this statement should be even stronger. Ventilation is *seldom* the best answer. Certainly, it is an expensive answer.

The prevailing assumption that the energy efficient "tight" buildings just naturally means less ventilation has lead to the belief that IAQ problems could be rectified by more ventilation. And for "ventilation" read higher energy bills.

The ASHRAE 62 standard is titled, "Ventilation for Acceptable Indoor Air Quality." To the uninitiated, that sounds like ventilation will deliver "acceptable" air. It may not. At the very least, the title sounds like the American Society of Heating, Refrigerating and Air-Conditioning Engineers, Inc. has given its blessing to ventilation as THE mitigating strategy.

As the various versions of ASHRAE 62 have been formulated over the years, the idea that much of the IAQ problems were created by, and could be cured by, ventilation has prevailed. Make no mistake, ASHRAE 62 has undoubtedly brought relief to many people, who would have otherwise suffered from "sick building syndrome." ASHRAE 62 guidance has been important during a time when it has been very difficult to determine what some of the pollutants were, what their levels of concentration were (or should be), and/or what their sources were. Increased ventilation has probably given relief to occupants during the time when we weren't quite sure what else to do.

Despite the impression left by the title of ASHRAE 62, *ventilation is not the preferred treatment for IAQ problems. It never has been!!* The Environmental Protection Agency has been telling us for years that the best mitigating strategy is control at the source.

As energy prices climb, we again suffer from reduced outside air in an effort to avoid the higher energy costs. With less outside air, we suddenly became more aware of the contaminants that had been there all along. Less outside air means greater concentrations. Since reduced ventilation has been a fairly standard energy conservation "remedy," it is not surprising that the knee-jerk response to the wide-spread dilemma of air quality problems has been to "open the windows."

In an effort to reduce energy costs, ESCOs need to be aware that increased ventilation has not necessarily been a solution for IAQ problems. Sometimes, in fact, it has made things worse.

LOSING GROUND

Below are a couple of instances where more ventilation may create greater IAQ problems. They represent circumstances where an ESCO may be pressured into implementing false "remedies."

Relative Humidity

Historically, when construction costs have exceeded the budget, the first cut, all too often, has been to remove the humidifier/dehumidifier equipment from the specs. Today, without those humidifiers or dehumidifiers, it is very hard to correct the negative impact increased ventilation can have on relative humidity.

Under conditions where the IAQ problems were virtually negligible, we increased ventilation and invited in all the IAQ problems associated with air that is too dry or too humid. With over 50 years of data on respiratory irritation, even illness, due to dry air, creating drier air by introducing more outside air during the winter in northern climates has not been the answer. With all we now know about microbiological problems and their relationship to humid air, creating more humid air through increased ventilation in places such as Florida has only aggravated those problems.

The Dilution Delusion

Mitigation by ventilation is dilution. But is dilution a desirable solution? Visualize for a moment all those airborne contaminants as a bright purple liquid flowing out of a pipe in an occupied area. Would an ESCO be apt to recommend treatment by just hosing it down each morning? Unlikely! Just because we can't see the air pollutants, does not mean they are any less of a problem. Or that dilution is an acceptable solution.

A critical health consideration rests on realizing that we may have not gotten rid of the problem by reducing the level of concentration. There is still much that is unknown about chronic low level exposure to some contaminants. A very real possibility exists that, in a couple of decades, studies will reveal that solution by dilution was nothing but delusion. A very serious delusion.

Determining the Value of Increased Outside Air

Using increased outdoor air as a mitigating IAQ strategy also makes several gargantuan assumptions. First, it assumes increased outside air is going to reach the occupants in the building. Milton Meckler found in a review of office buildings in the late 20th Century that over 50 percent of offices in the United States had ventilation designs that "short circuited" the air flow. This short-circuited air may have caused a nice breeze across the ceiling, but it did little for the occupants. ESCOs should be constantly alert as to where the diffusers are located and where the air leaves the room so the ventilation effectiveness can be assessed.

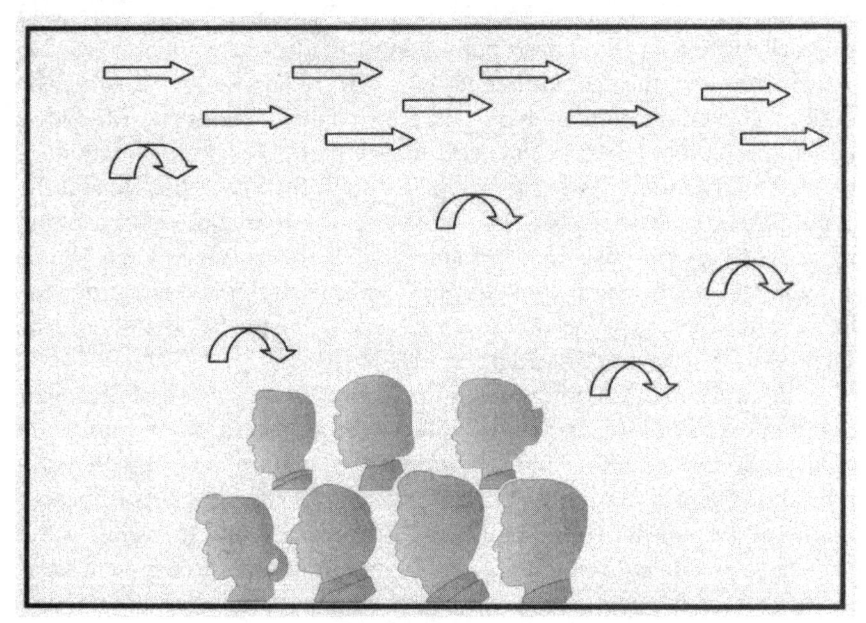

Figure 12-1. Ventilation Short Circuiting

Second, the outdoor air focus may have us bringing in outside air when re-circulated, cleaned air might be better. New outside air must be conditioned and circulated—often a costly measure. Using more outside air, which may be worse than inside air, can cost millions and millions of dollars. The fresh air fetish has too often overruled economics when effective filtration of re-circulated air could provide us the indoor air quality we need. We have also lost sight of the fact that increased outside air requires additional conditioning, which burns more fossil fuels and increases pollutant emissions.

ESCOs need to carefully review ASHRAE 62 recommendations for two reasons: 1) citing this source offers the best legal defense for any ASHRAE compliant actions taken; and 2) the standard does offer compliance through measures that do not increase ventilation (and energy use).

With regard to legal considerations, it is imperative that ESCOs only offer to look for sources of IAQ problems, or to suggest they might be able to improve the quality of indoor air. *ESCOs should NEVER **guarantee** they will resolve any IAQ problem, nor should they **guarantee** they will provide a certain level of indoor air quality.*

Historically, the needed amount of outside air intake has been gauged by the CO_2 concentration in the air. CO_2 measurements were used primarily to reflect the indoor population density, but were also used as a surrogate for other contaminants which were harder to measure. The initial "62" standards were based on odor control—the type of thinking which has its roots back in the Dark Ages when people took a bath once a year whether they needed it or not. Through the years, people and the people-based source of pollution—smoking—have made body odor and tobacco smoke the basis for ventilation needs. Unfortunately, using this as a basis for required air changes per hour has overlooked contaminants from a few other ages, such as the Renaissance Age, the Industrial Age, the Technological Age, and Information Age.

When ventilation mitigation is prescribed PER OCCUPANT, all the other pollutants, which have been introduced over the years, are not apt to be sufficiently treated. How many hundreds of new volatile organic compounds (VOCs) do we add to the list each year? As we "progress," people pollutants become less of a factor, and building materials, furnishings and "new, improved" equipment take on greater importance. European studies, referenced in ASHRAE 62 work, have shown that the building materials pollutant load today is much larger than we expected.

Occupancy-based ventilation levels are not sufficient if pollutant sources other than people dominate an area. Laser printers and copiers, as they operate, give off just as many pollutants whether there are 2 or 20 people in an office. Ventilation PER OCCUPANT simply won't meet such needs in low occupancy situations.

To confuse things even further, the National Institute of Occupational Safety and Health (NIOSH) labeled the illness from those energy efficient TIGHT buildings, "tight building syndrome." "Sick building syndrome" and "tight building syndrome" became synonymous.

Blaming tight buildings gave us charts like Figure 12-2; so we could compare those minuscule energy savings to the potentially HUGE personnel losses. Presenters, even from the US Department of Energy, compared energy savings to increased absenteeism and lost productivity, and stressed what an expensive idea energy efficiency was. Clearly, the hypothesis implied in the message was: *there is a direct correlation between energy efficiency and indoor air quality problems*.

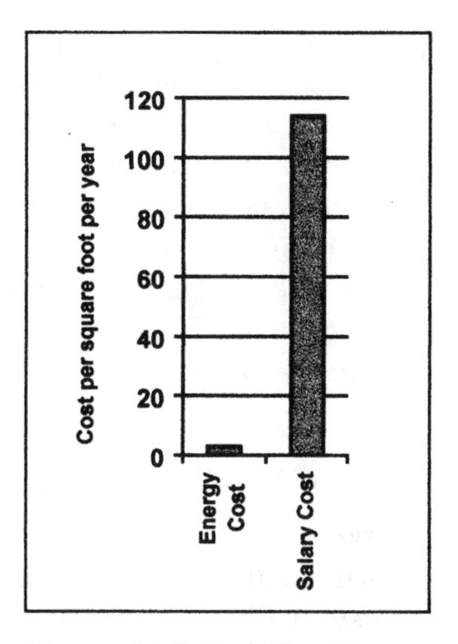

Figure 12-2. Relative Energy and Personnel Costs

To prove the hypothesis that lower energy bills make people sick, it would be necessary to show that all energy efficient buildings have poor air quality and lower productivity. With a little regression analysis, we ought to be able to build a straight line relationship… the more energy efficient a building becomes, the greater the absenteeism and the lost productivity.

Perhaps it is time to borrow the old procedure of taking the bird into the coal mine as a test for sufficient oxygen levels. Only now we'll take the birds into the energy efficient building!

The emphasis on tight buildings could lead an ESCO technician to believe that there was something really wonderful about those old leaky buildings. But those creaky, decrepit old leaky buildings were full of unconditioned, unfiltered, uncontrolled breezes. Truly an indoor air problem. The old line attributed to Casey Stengle is worth repeating, "Life ain't what it used to be and probably never was."

The unvarnished truth: "natural" air is not always desirable. Nor is there anything wrong with a tight building; provided that a tight building is well designed and well maintained. It is true that tight buildings more readily reveal professional errors as well as well as being less forgiving of poor maintenance. But a well-designed, well-maintained tight building can provide energy efficiency *and* quality indoor air.

Virtues of Ventilation

Ventilation definitely has its place in an indoor air quality program, a factor that ESCOs need to consider. Ventilation can serve as a mitigating strategy, when the contaminant or its source can't be determined, as an intermediate step until action can be taken, or when source mitigation strategies are simply too costly. Specific applications of ventilation; e.g., localized source control or sub-slab ventilation to control radon, can be valuable control measures.

The Real IAQ/Energy Efficiency Relationship

Research does reveal a relationship between IAQ and energy efficiency. For a long time, survey after survey has told us that when the utility bills start to climb, the first place owners, and many facility managers, looked to find money to pay those bills has been in the maintenance budget. This has been especially true for institutions on rigid budgets, such as public schools and hospitals. As the utility bills have gone up through the years, public institutions have progressively cut deeper and deeper into maintenance until the deferred maintenance bills are staggering.

The second relationship between IAQ and energy efficiency, can also be traced back to energy prices and maintenance. As energy prices climb, owners have bought more sophisticated energy efficient equipment. Unfortunately, O&M training to operate and maintain that equipment has not always kept up. Sometimes the training isn't offered when the equipment is installed. More often, there is a turn-over in O&M personnel and the new staff does not receive the necessary training.

Keeping these relationships in mind, it is really quite ironic and very sad, to learn that a majority of the IAQ problems are due to inadequate operations and maintenance. From the early days of IAQ research, the relationship between operations and maintenance and IAQ has been apparent. Table 12-1 reveals what we have known *for a long time*—that deferred maintenance has a very high price tag. Unfortunately, we have not recognized that a large piece of the price tag is paid for in occupant health problems. A review of the IAQ problems found by NIOSH, Honeywell and the Healthy Buildings Institute (HBI) at the beginning of the 1990s are depicted in Table 12-1. The labels are different; but the commonality of O&M-related problems is very apparent.

Table 12-1. Sources of IAQ Problems

Org.	NIOSH	HONEYWELL	HBI
Bldgs.	529	50	223
Yr.	1987	1989	1989
	Inadequate ventilation (52%)[a]	Operations & Maintenance (75%) - energy mgmt. - maintenance - changed loads	Poor ventilation - no fresh air (35%) - inadequate fresh air (64%) - distribution (46%)
	Inside contamination (17%)		
	Outside contamination (11%)	Design - ventilation/distribution (75%)	Poor filtration - low filter efficiency (57%)
	Microbiological contamination (5%)	- filtration (65%) - accessibility/drainage (60%)	- poor design (44%) - poor installation (13%)
	Building fabric contamination (3%)	Contaminants (60%) - chemical - thermal - biological	Contaminated systems - excessively dirty duct work (38%) - condensate trays (63%) - humidifiers (16%)

Source: *Managing Indoor Air Quality*, The Fairmont Press.

a. Percentages exceed 100% due to the multifactorial nature of IAQ problems.

THE MUTUAL GOAL OF IAQ AND ENERGY EFFICIENCY

Every ESCO, facility manager and design professional professes that it is their goal to provide owners with a facility that has an attractive, healthy, safe, and productive environment as *cost-effectively* as possible. If, indeed, that is the goal, then IAQ and energy efficiency are very compatible. They go hand in hand.

USING WHAT WE KNOW

80 percent of IAQ problems can usually be spotted with an educated eye and a walk-through of a facility. This walk-through might include some very basic measurements for temperature, humidity, CO_2, etc., but is definitely not a sophisticated, in-depth investigation. The other 20 percent of problem facilities require more specialized testing—often extensive testing. Even with exhaustive, expensive testing, about 10 percent of the problems remain unresolved. To summarize, 80 percent of the IAQ problems are detected through a relatively simple walk-through, 10 percent are resolved through pretty sophisticated, expensive testing and nearly 10 percent remain unresolved.

We also know that the single greatest cause for *in*efficient energy operations can be attributed to inadequate O&M. We also know that when the utility bills go up, owners typically find the money to pay those bills by cutting O&M, which creates even higher utility bills.

It is bitter irony that to save money to pay the utility bill, owners have cut operations and maintenance. Then, they have ended up with maintenance related IAQ problems—*and higher energy bills*. Then, the vicious cycle starts all over again… resulting in more cuts in the maintenance budget.

Fortunately, there is a positive side to this connection.

IAQ AND ENERGY EFFICIENCY: AN ESCO SOLUTION

The EE/IAQ relationship can be turned to an owner's advantage by an ESCO. If nearly 80 percent of our IAQ problems can be identified with a walk-through investigation; then pairing the walk through energy audit with it can set the stage for energy savings to pay for needed IAQ modifications.

The measures, of course, for energy efficiency and IAQ will vary, but with the right training, they can be spotted on the same walk through. An IAQ walk-through investigation can embrace the idea behind performance contracting: let future energy savings pay for the work. The walk through audit can identify the problems and can often find ways to finance the mitigation. This approach proves once and for all that IAQ and energy efficiency are compatible.

The combined IAQ/energy efficiency walk through can set the stage for the owner's relief from the specter of IAQ lawsuits by assuring them they have done what the "reasonable man" would do. In almost every case, the energy savings can make it possible to remedy the identified IAQ problems without taking a cut out of the budget. Offering a preliminary IAQ investigation as part of a scoping audit might also prove to be a strong marketing ploy for ESCOs.

Let's Not Kill the Messenger

There is one other relationship between IAQ and energy efficiency that deserves careful consideration. When ventilation and infiltration were reduced in the late 70s and early 80s, the net effect was to increase the concentration of some contaminants that were already there. Energy efficiency measures, in many instances, were the messengers, but they were seldom the *cause* of the problem. The energy measures only made us aware that we had pollutants in the air that could be injurious to our health. We learned that the sources of those pollutants should be removed or contained whenever possible. Energy efficiency gave us the message. We need not take a page from the king and shoot the messenger.

Mutually Compatible Measures

As the cost of energy goes up, once again IAQ and energy efficiency are apt to be at logger heads. This does not have to be the case. If IAQ leaders persist in attributing IAQ problems to energy efficient buildings, thus relying on more and more outside air for the answer, we lose. For cost-conscious owners climbing energy costs may outweigh most IAQ concerns. Only if government steps in by regulating expensive, and often unnecessary, outside air requirements or a court orders such measures, are owners apt to serve IAQ needs in the face of escalating energy prices. Under such conditions, the money could still be wasted and IAQ needs may not be served.

Environmental concerns, higher energy prices, national security issues and the unnecessary waste of limited energy resources make increased ventilation, at the very least, a costly answer. Sustainable development requires a collective effort to work together for a quality indoor environment, energy efficiency AND a quality outdoor environment. ESCOs are in an excellent position to bring such resolution to light.

The ultimate goal is to produce a comfortable, productive indoor environment as cost-effectively as possible. That puts energy efficiency and indoor air quality on the same side. Good managers and ESCOs need to have command of both if they are to do their jobs effectively.

Section III

Performance Contracting: The Next Generation

We shall not end dependence on imported oil, nor... end dependence on the volatile Middle East, with all the political and economic consequences that flow from that reality. Instead of energy security, we all have to acknowledge, and to live with, various degrees of insecurity.

—James Schlesinger, 2005

Carved in stone on our National Archives is the Shakespeare quote, "What is past is prologue." But it has apparently not been carved into our consciousness. In 1973-74, when our oil imports were 26 percent, OPEC gained the upper hand with an oil embargo, which cut our supplies, sowed economic turmoil and a global recession. Now our imports are nearly 60 percent, the energy weapon others can yield over us has grown to gigantic proportions. We have had over thirty years to learn, and we have only made it worse.

As Russia briefly tried to cut off gas to Ukraine and indirectly across Europe in early 2006, the world saw the energy weapon and its repercussions once more emerge. And we did not even bat an eye.

Trying to capture how the next generation of ESCOs will function in such an insecure energy environment is exceedingly difficult. All the signs, however, suggest there will be considerable upheaval in our energy world in the years to come. It looks like it could be a wild ride.

Strong indicators suggest that energy prices will trend up, availability will become as critical to many owners as price, renewables will gain a stronger market position, and conservation will become a dominant energy strategy. These predictions, of course, could be hugely impacted by unknowns, such as the timing as to when, if ever, we wake up to the need to secure more domestic resources. In the meantime, as Former Energy Secretary Schlesinger tells us, we can expect various degrees of energy insecurity and uncertainty.

It is clear that our energy crystal ball is very fuzzy—and getting progressively harder to read. In this section, therefore, we will focus on the immediate horizon for performance contracting.

In the first edition of this book, we stated, " it seems pretty safe to predict the services offered by energy service companies and utilities will change; customers options for energy services will broaden; and options for performance contracting will expand."

Those predictions have proved to be true. The ride is still wild and even more unpredictable. Now it is time to look at how those changes have manifested themselves and how they will continue to evolve. Chapter 13 takes a new look at the currently emerging options to expand the ESCO model. Special attention is paid to how ESCOs can fit into the growing trend towards outsourcing and privatization. Despite the complex supply picture, or maybe because of it, attention is paid to integrated solutions and chauffage.

We also stated in the introduction to Section III of our first edition, "we certainly cannot address the future without acknowledging the growing presence of the Internet, and the changes we have witnessed in how we communicate and do business." In all honesty, we did not expect web-based activity to come so far so soon and become so pervasive in the business world. We are incredibly pleased that three gentlemen, who understand this growing Internet world, have put their heads together to give us an excellent view of web-based energy data acquisition and analysis in Chapter 15. Paul J. Allen, David C. Green and Jim Lewis have offered us an exciting glimpse into ways this new communications era can help ESCOs serve their clients more effectively and efficiently. Paul is the Chief Energy Management Engineer at Reedy Creek Energy Services, a division of Walt Disney World Co., and as such is responsible for the development and implementation of energy conservation projects throughout the Walt Disney World Resort. Paul was inducted into the Association of Energy Engineers Hall of Fame in 2003. David has been the president of his own consulting company, Green Management Services, Inc. since 1994 and has extensive experience in Intranet/Internet technology and database queries. David is also a Lieutenant Colonel in the Illinois Army National Guard and has 18 years of military service. Jim is the CEO and co-founder of Obvius, LLC. He was previously the founder and president of Veris Industries, a supplier of current and power sensing products to BAS manufacturers and building owners. He served as a Branch Manager of Honeywell and has extensive experience in integrating existing meters and sensing technologies as well as developing innovative products for dynamic markets. Their combined expertise and experience provides ESCOs, and indeed all of us, with a great introduction into web-based procedures.

Chapter 14 breaks with the trend and does not build on the previous edition. Instead, we have been given a challenging and provocative look at how vendors and ESCOs can work together for their mutual benefit. Michael Gibson has worked in the energy service industry for twenty years. Over the last ten years, Michael built a Florida-based vertically integrated energy service company. He is president of BGA, Inc. a specialist in project development and program consulting. Brian Todd has been involved in new Product Technology businesses for 24 years, with Westinghouse Electric Corp., Honeywell International and Dialight, LLC. Brian is currently Vice President for Dialight. Michael and Brian have found a way to forge an effective vendor/ESCO partnership, and share their insights in "Productive Vendor/ESCO Relations." Their approach provides a good model for critically examining many aspects of the ways ESCOs do business and to determine if similar approaches might be used in other aspects of the ESCO business.

Chapter 13

Expanding the ESCO Model

As performance contracting developed in the US, its focus was initially on schools, hospitals and municipalities. These creditworthy, cash-poor customers made ideal performance contracting candidates. Their perennial budget shortfalls, consistent and annual energy usage patterns made a good match for the embryonic ESCO industry. The public sector served as an excellent demonstration of the concept, for universal needs were met with proven solutions and millions of dollars saved.

By the 1990s, the concept of performance contracting seemed to have incubated long enough, gained broader appeal, and was soon being offered to virtually every market segment from supermarkets to refineries. The prevailing theory was, and still is, valid for most energy users; i.e., a project, which could fund itself from energy and/or operational savings within a reasonable time period was an economically sound proposition. However, this theory, which appealed to the local school board, has been

often rebuffed in the industrial sector.* As discussed in Chapter 1, "energy" is all too often not on management's agenda.

Options That Expand The Model

New markets as well as new and expanded services offered ESCOs many options for expanding the performance contracting model.

Expressions become trite because they hit a responsive cord. Such is the case, when we urge those involved in developing ESCO opportunities to "think outside the box." Consider, for example, a performance contract where energy consumption actually increases, but the concept still works. It is time to examine some of the commercial and industrial sectors' unique needs and see how our industry might serve them more effectively.

THE INDUSTRIAL OPPORTUNITY

As noted in Chapter 1, industry is a hard sell. If we can develop this market segment, it should help us effectively penetrate many other markets.

To meet industrial needs, it has been necessary for ESCOs to expand the organizational performance contracting model and modify it to more effectively meet a different set of consumer needs. Project designers have been challenged to go beyond the traditional school facility model and examine commercial and industrial (C&I) procedures and processes and determine how performance contracting might offer a better fit.

There are so many variations of products, processes, energy costs, balance sheets, and even market stability that we can only reference a "typical" industrial opportunity. Although the need set for any particular school district usually matches that of other districts, industrial needs vary far more widely and require a:

a) much more demanding prequalification process; and
b) projects that more effectively meet the unique needs of each industrial customer.

*For the purpose of this chapter the term "industrial" refers to any processing facilities, such as an aluminum smelter, chemical processor, or automobile manufacturer.

Traditional performance contracting, so efficiently applied to the predictable requirements of the institutional customer, has often missed the mark in industry. This has largely been due to the fact that the industrial customer's desired result is significantly different from that of the institutional client. Where a typical school district might have a need to upgrade the learning environment through better lighting and temperature control, the industrial management strives to increase shareholder wealth.

Although any reasonable cost reduction measures are welcome, the implementation value is always weighed by management against the risk of lost production or the potential for increased production. If the projected cost of lost production or reduced yield even speculatively outweighs the value of energy cost savings, the project will be rejected. Negative impact on output quality will almost certainly cause a performance contracting proposal to be shelved.

Processes that operate 24 hours per day have the potential to benefit the most from energy savings. However, projects are most at risk should implementation cause process interruptions. In such circumstances, downtime needs to be factored into project costs. This factor needs to be applied to commercial customers as well. For example, the cost of space in Hong Kong is so exceedingly high that retail outlets cannot afford any day time closure; so any retrofit work in a retail store must be done at night.

EE measures must also be weighed as to their impact on the work environment. For example, installing energy efficient high pressure sodium lamps with their characteristic orange light and poor CRI might reduce energy consumption, but the quality of color sensitive processes such as printing or dyeing would suffer.

After a manufacturer's fixed costs are absorbed, ongoing production becomes considerably more profitable and any energy solution that might retard production after break-even is totally unacceptable. So, the term "risk" in an industrial performance contract becomes the risk of lost profits—a markedly different perspective than the political risk a poorly negotiated performance contract presents in the municipal market.

Most "for profit" corporations ignore the difference a self-funding project offers and evaluate a performance contract as critically, methodically and stringently as any major equipment purchase agreement. Industrial managers typically view energy efficiency work as an expense; not an investment. When top management does consider it as an investment, it compares EE work as any other product acquisition using such yardsticks as IRRs, hurdle rates, and ROIs. CEOs and CFOs seldom recognize that EE can be totally self-funded, which should require a totally different yardstick. For more complete discussion of this issue, the reader is referred back to Chapter 1.

To the manufacturer, production is king. Energy costs in such process industries as chemical, aluminum and cement can be as much as 30 percent of the resources required for a unit of product while it can be as low as 2-5 percent in some industries. When energy constitutes a small part of the budget, its relative significance to a company is diminished. Labor and materials may account for the largest cost elements and are often the focus of industrial engineers and process designers when cost cutting efforts are under way.

A Performance Contract that Increases Energy Consumption

Nothing shakes our preconceived ideas of what a performance contract should be than to point out that the procedure can work in situations where the energy consumption is actually increased. If increasing energy consumption improves production and subsequent profits, an industrial performance contracting project can be desirable. A manufacturer of automobile carpets produced from synthetic materials cured one side of the goods in a single pass through a natural gas-fired oven, rewound the carpet, and repeated the pass for the opposite side. A conversion to an electric infrared oven that cured both sides simultaneously increased the cost of energy marginally but doubled the production of the curing process. The resulting simple payback period for the project was just 6 months.

The above example provides a clear-cut example of ways performance contracting in the private sector can differ from the public sector. The public sector does not gain the benefit of this upside potential since improved learning does not generate additional tax income to pay for the increased costs of enhancing the learning environment. The industrial market allows an ESCO to propose solutions that make prudent business sense, whether it reduces energy consumption or not.

A guarantee of performance is an integral part of the traditional performance contracting process but, contrary to the highly sensitive political environments of public institutions, the industrial customer can accept a broader range of performance goals. Commercial and industrial (C&I) management typically does not need the same reassurances that public administrators do. For example, the C&I people are less apt to find overwhelming value in an intensive long term measurement & verification (M&V) process. In contrast to the public sector, risk taking is a normal aspect of the business climate and the C&I customer is not likely to seek insurance or reassurance on every business decision. However, industry typically offers ESCOs a prudent and savvy customer, one who prefers to protect his investments by dealing only with companies that have demonstrated capabilities, long-term stability and substantial financial resources.

THE PERFORMANCE CONTRACTING TIME HORIZON

Yet another issue that impacts the desirability of performance contracting outside of the traditional markets is the contract period. In areas of the country where electricity or natural gas costs are relatively low, a self-funded project may also require lengthy payback periods to support the needed positive project cash flow. The savings and service opportunities as well as the price of energy, which dictates payback terms, influence contract length.

A municipality may expect to own and occupy its buildings for an indefinite period of time and, therefore, can accept longer projects. The "time horizon" tends to be much shorter in industry due to economic pressures and uncertainties. Management, which lives with a "Fourth Quarter" mentality, has a hard time thinking about even a five-year contract term. Product obsolescence, market shifts, imports, mergers, and a host of other uncertainties can leave a business liable for payments for energy efficiency improvements that outlive the company's full utilization of the facility. Accordingly, this customer expects a quick simple payback and a contract period in the 3-5 year range as opposed to the 8-12 year range acceptable to most public institutions. In low to moderate energy cost areas, savings from energy costs alone will not be adequate to produce a performance contract period in the target range—a primary reason why energy use reduction alone is not always the answer.

THE ALL-PURPOSE ESCO

Traditional performance contracting usually has limited its scope to conventional energy efficiency and conservation measures on the demand side of the energy equation. Conditions are prime for a growing generation of "integrated energy service companies," which can offer energy efficiency, water and waste water treatment efficiencies, supply acquisition management, asset monetization and energy accounting services. Innovative financing can allow an industrial customer to recognize positive cash flow throughout the contract. "One stop shopping" for all of these needs reduces purchasing costs and can generate immediate earnings.

Some ESCO's have even offered an inducement of an initial cash payment at signing equal to the discounted value of the project's guaranteed savings cash flow stream. However, any bold action of this nature increases the ESCO's financial risk and can translate into additional cost to the customer.

SUPPLY ACQUISITION

When energy supplies become a political tool—as Russia tried with Ukraine in early 2006, it raises uncertainties and often complicates the purchase. Owners are apt to need more acquisition assistance. This could present a business opportunity for ESCOs; however, it will probably cause them to incur more risks.

In order to reduce an energy efficiency proposal to an acceptable contract period, savings may need to be accumulated from other sources. Supply acquisition management is an increasingly important tool. As electricity delivery becomes less regulated and follows the pattern of unregulated natural gas supplies, an optimum purchasing strategy will grow and can become the best value for the customer. Major national industrial and commercial companies with dozens to thousands of locations can "aggregate" their energy supply purchases to obtain the most favorable rates and reliable service. A new generation of ESCOs with a broader range of services will emerge, highly qualified to participate in national retail energy markets.

Any assertion that ESCOs must choose to participate in either the supply or energy efficiency industries, since no company can excel in both, is a protectionist theme for myopic first generation ESCOs.

The second generation ESCO, once described as a "Super ESCO," will guarantee lower costs based on its ability to purchase supplies more effectively, particularly where the ESCO can exercise fuel switching on short notice. A manufacturing process that is designed to use multiple fuels allows the ESCO with a staffed and active trading floor to take advantage of market volatility and execute spot purchases that are exceptionally economical for the end user. While a few industrials have this staff and capability in their organizations, there are many that would be pleased to outsource such an activity.

Performance contracts, where the customer and ESCO both participate in actual savings, are increasingly being used as supply acquisition instruments. Just as the magnitude of an energy efficiency contract is determined by the extent of the customer's own previous efforts, the amount of guaranteed supply savings will be determined by the customer's existing skills at purchasing fuels and energy. As electricity markets open up, the skills required to purchase energy will escalate, leaving countless millions of dollars to be captured or lost to indecision.

INTEGRATED SOLUTIONS

Waiting in the wings for the more aggressive ESCOs is an opportunity to provide the client both energy demand and supply services. Offering supply services or pricing assistance plus demand efficiency is an integrated solution (IS)—integrating services from both sides of the meter for the one client. The supply side may include DG, CHP, contracted acquisition, or one might grow a power plant.

This can be presented as two separate services for the same client; or, it can be a blended offering when an ESCO offers conditioned space at a specified price per square foot or square meter. This more complete blending of services is referred to as chauffage.

Chauffage can have a surprisingly simple contract. The legal arrangements can focus on specified ranges of workplace conditions, such as temperature, lighting levels, and air changes per hour with stipulated payment by square or cubic area. Provisions for annual price adjustments, much like the fuel adjustment clause in the local utility bill, can be included in the contract.

When an ESCO adds the supply acquisition to its repertoire (so it can offer IS or chauffage), supply expertise must be added to its stable of talents. The tools and practices of energy trading and/or option hedging become critical. Today's energy markets and hedging techniques must include fixed price concepts (price curves, basis, etc.) and such tools as forwards, futures and swaps. All this expertise can be resident in the ESCO staff, trained or acquired. The option also exists to form an alliance with a firm with supply expertise.

It should also be noted that in the wake of the Enron debacle, there have been some dramatic changes in contracting. The laws, regulations and contracts related to energy trading have become a specialize field requiring constant vigilance as FERC and states seek to control action in the market place.

An earlier caution is worth repeating: ESCOs looking to pursue supply acquisition services and/or integrated solutions/chauffage, will need to keep a closer eye on the international scene. Should a country choose to exercise political power through the energy weapon, an ESCO's supply scenario could change within hours. Protective contract language for such events is strongly recommended.

CUTTING EDGE TECHNOLOGY

Nearly every ESCO has had a client that requested, even insisted upon, some new cutting edge technology. Unfortunately, "cutting edge" generally translates to new and untried. It is, therefore, extremely risky for an ESCO to guarantee the performance of essentially untested equipment. Issues like the level of savings over time, maintenance requirements, repair and replacement needs all represent unknowns. It is not long before the ESCO begins to feel like it's performing a high wire act, as it tries to satisfy the owner and still limit its risks. The answer to such pressure is to give the client what they want, but to keep this cutting edge equipment outside the guarantee parameters. Such an approach allows the ESCO to broaden its offering and be more flexible in meeting clients' needs without incurring greater risks.

STRUCTURING A BUNDLED SOLUTION

Once an ESCO has identified opportunities to reduce energy consumption, increase production through process modifications, and/or purchase energy more cost effectively, it may help to broaden the scope even further. Consider the textile manufacturer that had been paying a flat fee of $100,000 per year to the local municipality to treat effluent from his fabric dyeing operation. At the time the treatment plant was built by the town, this cash flow was substantial and contributed to reducing the town's debt; but, as the years passed, the costs of treating the plant's waste exceeded the payment. In a bold confrontation with its largest employer, the city increased the plant's sewage charge six fold, to $600,000, to match prevailing rates in the area. In addition, the town required that the plant install $300,000 of flow metering equipment to accurately measure the actual flow. Unwilling to accept the new charges and determined not to divert capital from new production equipment, the company turned to an energy services company for a solution.

The ESCO explored many alternatives that included recycling the wastewater, building an on-site treatment facility, and pumping the waste to another plant three miles distant that had unused capacity available at a low cost. Recycling created an unacceptable risk of defective product caused by potential failure of the filtering equipment that removed colored dyes from the water. Also EPA permitting requirements for on-site treatment were prohibitive. However, property rights between the plants were owned by both companies and the cost of pumps and piping would allow recovery of the investment in five years. By installing heat recovery equipment on the 190 degree F wastewater stream and using the recovered energy to preheat boiler make-up water, the ESCO showed additional savings that reduced the proposed contract period to four years. Converting the steam boiler to dual fuel and providing the lowest cost energy at any time of year at guaranteed maximum rates produced supply savings that further reduced the contract period to nearly three years. The ESCO provided a turnkey project including funding, design, project management, construction, and maintenance—and met the desired time line.

By applying the familiar concept of the self-funding proposition and structuring it within the fundamental industrial time horizon and production risk constraints, an ESCO can meet a company's business needs as well as its energy needs within the performance contract model. Rather than considering performance contracting as merely a financial vehicle for enabling the sale of the proposer's products, systems and/or services, it can be viewed as a powerful tool to improve production and earnings. Further, an ESCO can use energy savings to buy down the costs of a more comprehensive project.

OUTSOURCING AND PRIVATIZATION

"Asset monetization" in which the ESCO purchases existing assets, is yet another offspring of the performance contracting offering. A new focus on profitable core business, often coupled with corporate executive incentive plans that reward high (ROA's), encourages decision-makers to dispose of assets that can be operated and maintained more efficiently and reliably by third parties. Steam, chilled water, and compressed air are typical processes that are essential to production but excellent candidates for an outsourcing agreement. The objective of the facility owner is to reduce and control his costs associated with using these media and processes.

The ESCO can calculate the facility's current cost of consumption of these commodities with a reasonable degree of accuracy. If these commodities can be delivered cost-effectively for less than the established baseline, the fundamental conditions for performance contracting can be met. Savings, which will allow the ESCO to earn a profit and the owner to recognize a lower cost, come from several sources, including:

- improved operational efficiency resulting from the ESCO's superior engineering resources;
- lower cost of fuel and/or electrical supplies; and/or
- lower cost of maintenance after equipment is upgraded.

The driving force behind the ability of the ESCO to engineer real savings is the fact that most facilities have grown in stages over the years so that plant systems are an assemblage of add-ons and quick fixes. Few systems have been designed for their current capacity, resulting in major inefficiencies from undersized distribution conduits, piping conflicts, leaks, and other expensive deficiencies.

Another source of savings results from the ESCO acquiring capital at a rate lower than the customer's required rate of return, allowing the customer to reinvest the proceeds from the sales at a higher rate. A variation on this offering is to determine the current energy/capital consumption of the customer's product and offer a lower, guaranteed maximum cost per unit of production for the essential commodities on an outsourced basis.

As attractive as these propositions may sound, let's restate that for industry, production is king and any real or perceived threat to production from a transaction can quickly shelve an otherwise exciting opportunity. Accordingly, only ESCOs with a solid reputation for performance, specific outsourcing experience, and substantial access to capital will find this type of venture to be fertile ground.

In summary, traditional performance contracting is transferable to the "industrial" sector and other markets if the ESCO can gain a clear understanding of the customer's specific business needs, not just energy needs, and offer a broad range of solutions that meet those requirements. Energy alone is seldom the answer for a customer whose driving force is accelerated production, lower operating costs, and, most important, increased earnings.

Chapter 14

Productive Vendor/ESCO Relations

Historically, the vendor has typically sold equipment—only equipment. Unfortunately, equipment by itself is a commodity, which becomes primarily differentiated by price. To be successful in the market place, vendors increasingly need to offer value-added services. Energy service companies (ESCOs) can provide the vendor quality service to enhance the product while, at the same time, offering an effective marketing channel.

Direct purchase is typically a one shot deal, while an ESCO provides services and customer satisfaction *over time*. Through the ESCO, the vendor gains long term exposure to the customer and inherent resale opportunities.

The ESCO guarantees results and an important part of those results is based on reliable, quality equipment. A manufacturer, who will stand by its product over time, reduces ESCO risk. The vendor can also deliver potential customers to the ESCO.

Today, the traditional delivery of non-energy service construction projects is giving way around the world to the more efficient design build process. So, it is time to rethink the traditional energy service delivery mechanism and take the design build concept to the next level: the guaranteed results of performance contracting. In this result-oriented approach, energy efficiency savings buy down the equipment costs thus offering the customer a blended product/service price.

The energy service industry is all about providing best value. In keeping with this concept, this chapter will describe teaming opportunities between an energy service provider and an equipment, i.e. technology, manufacturer to deliver a higher-value energy service program to the owner through alternative delivery mechanisms.

This chapter will describe effective working relation opportunities as they relate to the energy service industry, including:

- how the alternative delivery mechanisms work with its turnkey design build,
- the way the approach offers advantages through early project development collaboration,
- the manner in which industry goals between manufacturers and ESCO's align, and
- how the approach is validated through improvement in overall cost efficiency.

Owner's Perspective for ESCO Services

It is well established that facility owners find value in ESCO services because these services improve efficiencies, improve comfort conditions, reduce maintenance costs and provide new equipment in a cost effective package by monetizing operational savings streams.

In the traditional sense, the energy service industry is the application of technologies that realize operational cost reduction. By leveraging these operational cost reductions, one benefit to the owner is that these desirable technologies, which would normally be capital expenditures, can be paid for through savings. The broader benefit is reduced operating costs paid for without capital budget expenditure.

In short, while services are the cornerstone of the ESCO industry, innovative technologies give the ESCOs improving opportunities to drive their projects.

To emphasize this point just consider the Ernest Orlando Lawrence Berkley (LBL) National Laboratory report published in March 2005, titled "Public and Institutional Markets for ESCO Services: Comparing Programs, Practices and Performances." This report evaluated over 1,634 energy service projects. All 1,634 projects could be categorized in just 34 broad categories, of which 33 represented application of technologies; clearly, an extraordinary impact.

Moreover, since there are so many technologies, fitting various applications, it is reasonable to expect emerging technology cannot be a core competency of ESCOs. Considering the critical role equipment plays for the energy service industry one would assume that ESCOs and vendors would, by now, have established strong teaming relations to optimize the delivery of such technologies. Yet, from an energy service industry standpoint, examples of such teaming are uncommon.

HISTORICAL ESCO DELIVERY

An owner generally follows an ESCO's developed project recommendations as well as the ESCO's preferred method of project delivery. For example, BGA, Inc., which has developed nearly $400 million in energy service projects, implemented close to 95% of the measures it recommended to its clients. Moreover, in no case did the owner modify the recommended project delivery approach.

Despite the project process flexibility of the owner and despite equipment applications being critical to the ESCO projects, the dynamics between vendor and ESCO, for the most part, have not been a project teaming approach. Though the benefits of a vendor and ESCO teaming arrangement are abundant, most energy service projects procure their critical equipment through traditional cost-based procurement.

INITIAL BARRIERS TO VENDOR AND ESCO TEAMING

Several barriers discourage such a teaming arrangement, including:

1. The ESCO project elements and project process
2. Project margin pressure
3. The actual equipment buyer
4. Project timing
5. Product warranty vs. PC contract savings period

The ESCO Project Elements and Project Process

An ESCO selected to deliver an energy efficiency project to a client must integrate as many as a dozen project elements, of which equipment selection and procurement is just one. To be sure, in addition to procuring the technologies, an ESCO offers a valuable range of services, including:

1. Project scope development
2. Savings and cost estimates
3. The negotiation of a workable energy service contract,
4. Construction management,
5. Commissioning the project,
6. Equipment maintenance,

7. O&M staff training,
8. Securing or arranging the project financing,
9. Savings measurement and verification to validate performance guarantees.

Moreover, each project element involves many steps. In essence, the ESCO project process involves at least 150 steps to deliver a final project. In light of all the project elements, it is no wonder product procurement becomes more of a project implementation than the orchestrated involvement of a strategic team member.

Project Margin Pressure

It is expensive for an ESCO to be an ESCO.

Much of the early project work is done at risk. Some of the project revenue is even spent before the owner selects the ESCO. It is not uncommon for an ESCO to burn through 2% of project revenue potential to just get selected. The 2% revenue potential equals approximately 6-8% of the gross project profit potential. And, when the owner does not select an ESCO, that ESCO must recoup those sunk costs on future projects.

Then, after selection, the real costs begin. Energy service projects are in large part developed at risk. Even if the owner pays for the development study, that fee rarely covers all the ESCO's costs. Additionally, an energy service project fee must include a contingency for project savings, performance risk, pay for the contractor, and procurement of the equipment from the vendor. Finally, despite all the costs that go into an energy service project, there is a ceiling to the project fee driven by the project savings. Project savings pay back the initial project fee. Since the payback generally needs to fall within the owner's financial objectives, or the statutory rules if a public entity, there is apt to be a ceiling to the project fee.

So, the ESCO is pressured on both financial ends of the project, its fee potential and its costs. This situation can lead to tight project gross profit margins—the fuel that feeds the ESCO project development burn rate. One option available to the ESCO is to buy down its project costs through a competitive bid process.

Because of this, by the time the project is developed to contract terms, and because about 30% of the project's cost is for the procurement of products, the ESCO is apt to look at equipment as a significant project cost, rather than the vehicle for the savings, which drive the project. As such, the equipment supplier can easily find itself in a commodity relationship with the ESCO. Hence, the vendor is now in the price-differentiating world for a significant part of the project, not the performance world. This is definitely not the most beneficial environment for fostering strong vendor-ESCO teaming.

The Actual Technology Buyer

Unfortunately, the above discussion may be the best-case scenario between the ESCO and the vendor. In the worst case, the problem can be especially bad for the project, both in the way of project quality and project performance, which are both project warranties, when the contractor hired by the ESCO tries to value-engineer the project and buys a less efficient, or lower quality, equipment. Without tight ESCO oversight, the vendor becomes one step removed from the ESCO. In a worse price-driven scenario, where the contractor also needs to preserve its gross profit margin, the trade off of long-term value for project profit margin can, and will, occur.

This darker side to the price-driven spiral is further impacted when contractor, as in most cases, is not involved in the project development effort. In such instances, the contractor cannot have the same sensitivity to the product efficiency and product warranty that the ESCO needs to prove the project.

As mentioned, this scenario can and will happen. BGA has paid out only 0.03 percent in energy efficiency guarantee shortfall and 0.60 percent in premature equipment warranty failure. While the aggregate payout of 0.63 percent is still much less than the 2.5 percent payout experience of the industry, it is still significant.

Having an engaged equipment vendor will go a long way to mitigate both the energy performance and the equipment warranty experience of the performance contract project.

Project Timing

The high project selection and project development burn rates, coupled with gross profit pressure, are now exacerbated by the long project development and sales cycle, which makes the ESCO's monthly revenue and income stream very lumpy. Once a project gets approved for implementation, the ESCO is under pressure to procure the equipment and install that equipment—quickly.

The need for quick delivery continues to push the product procurement to the role of the contractor, which will allow the project's quickest delivery path. It could be argued that the quickest delivery is to involve the vendor in the development process, so the product will be readily available upon work authorization. But without this early involvement, the next quickest method is to procure the product through the contractor with adherence to the ESCO's specifications.

Product Warranty vs. PC Contract Term

The ESCO is burdened with balancing project simple payback, contract length and the product warranty term. It is a certainty these three terms rarely align. The product warranty period is generally less than the payback period, which is less than the contract term. For example, if statutes limit a contract term to a 10-year contract, the project must be less than a 7-year payback period (pre-finance) and the product warranty will be less, at perhaps a 5-year warranty period.

Yet, it is the warranty period that helps the project economics. So, even while the ESCO is trying to find the lowest product costs, the ESCO is also trying to bridge the gap between warranty period and contract duration. One way the ESCO can bridge this gap is by securing a longer warranty period from the vendor. But, this extended warranty is not consistent with the price-differentiating environment the ESCO has established with the vendor.

VENDOR TRADITIONAL ROLE

Typically, the ESCO looks at vendors as equipment providers and assesses them by project cost. To illustrate the point, let's consider the four marketing channels leveraged by Dialight, Inc.:

1. Specialty distributors
2. Product representatives
3. Large equipment OEM's
4. ESCO's

Of the four, the ESCO market channel is the one channel that pushes price differentiation the most. With ESCO's, only 20% of Dialight's LED sales are negotiated price and terms, while 80% of sales are bid based. For comparison, sales through distributors and representatives are 60% negotiated price and terms; and in the case of OEM's, nearly 99% of sales are negotiated price and terms.

Clearly, as shown in Figure 14-1, the ESCO market channel is least differentiated by price and terms; and most differentiated by price. It is ironic that the one channel that could promote best overall value, not lowest first cost, tends to treat the technology driving its project's operational savings as a commodity. Of course, as described earlier there are project margins and project delivery drivers that have set this approach as the ESCO's de-facto product procurement method. Yet, there are clear opportunities available if a different project delivery approach, which is not just price-based, is considered.

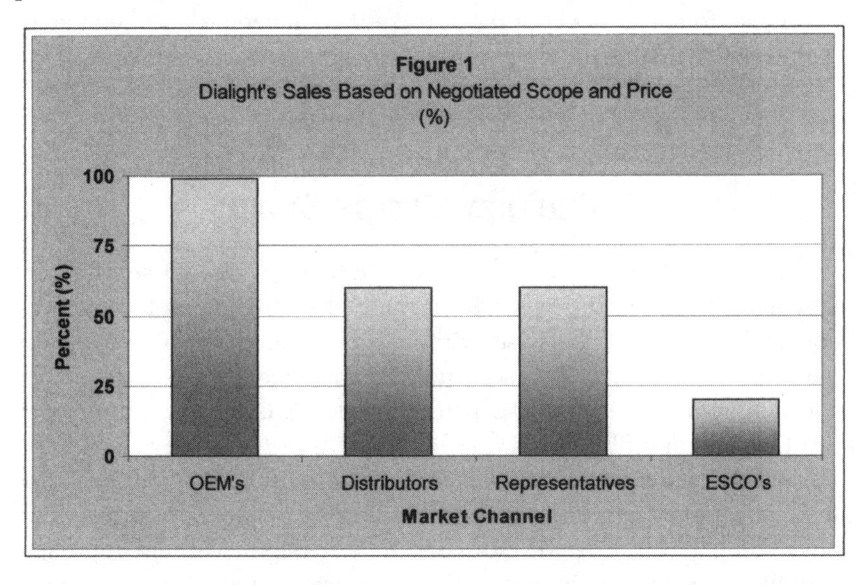

Figure 14-1. Scope and Price of Marketing Channels for Dialight

BRIDGING THE VENDOR-ESCO APPROACHES

Facility owners have increasingly accepted design-build delivery as a viable alternative for new construction and retrofit projects over design-*bid*-build. In fact, in 2001, design-build accounted for 35% of projects delivered. Further, it is anticipated that within this decade design-build acceptance will grow to represent 50% of projects delivered.

Considering design-bid-build had been the standard form of project delivery for well over 100 years, the last decade has seen a radical shift towards design-build.

Key reasons contributing to design-build acceptance are that design-build:

1. Is quicker than design-bid-build,
2. Offers contractor insight upfront in the project scope development,
3. Provides the owner a single point of responsibility, and
4. Offers the owner a mechanism to select the design-build firm that offers the best value, as opposed to the lowest price.

These four reasons sound a lot like the challenges and benefits of energy service projects.

Rethinking Delivery Mechanism Offered by Design Build

Ironically, from a facility owner's perspective, ESCO services are really just design-build projects turbocharged with a guarantee and a financing mechanism. Otherwise, without the guarantee and financing elements, ESCO projects have a scoping element, a single point contract responsibility for the design and construction elements, as are common with any other design-build project.

But, at least from a product standpoint, the ESCO rarely approaches the project as a design-build exercise. To be sure, most ESCO projects are delivered by the ESCO in the more traditional design-bid-build approach. With this approach, the implementing contractor usually negotiates with the product provider. This approach puts the equipment provider further away from the ESCO and aggravates the commodity position of the critical technology element.

Since the facility owner already perceives the energy service project as a design-build process, the door is open for an ESCO to re-evaluate its delivery process and consider a design-build approach using the equipment provider as a key project partner.

TEAMING OPPORTUNITIES

To be most effective, the ESCO needs to select its vendor partner at the beginning of the project. An equipment provider can offer considerable insight in project development. This collaboration can create a more complete development scope, share an alignment of risks, pool sales resources, and protect *project* gross profit without sacrificing *product* gross profit margin.

Vendor Pulls in the ESCO

A vendor with a strong product represents a significant market channel opportunity for the ESCO.

Most equipment providers have front-line sales. These sales resources touch many potential buyers, including facility owners interested in procuring such technologies, but who have a capital budget shortfall. This resource would be a perfect vendor-ESCO teaming opportunity, as the ESCO specializes in monetizing the savings stream the equipment offers. The ESCO benefits by having this additional marketing channel

The overall benefit of the ESCO/vendor collaboration allowing for these types of projects to be introduced into new markets and to new clients cannot be overstated. More products are sold to owners through traditional procurement methods than through energy services. To emphasize the point, within Dialight at most only 7% of product sales are to ESCO's. Around a quarter of the remaining 93% of sales are directly to the 200 to 300 potential direct customers Dialight meets with each year. The other three-quarters of the 93% of sales is accomplished through the other three well-established market channels.

ADVANTAGES

Before noting further advantages, it will help to first summarize the three vendor-ESCO teaming advantages that have been identified. They are:

a) The design-build process is quicker, allows earlier collaboration in the critical project development phase, and mitigates positioning the contractor between the ESCO and the technology provider. In the purest form, both the contractor and the vendor become part of the design-build team.

b) Negotiated quotes open the door to considering other factors that bring value beyond lowest first price. Examples of other values include product warranty and product efficiency assurances.

c) Both parties enjoy expanded market channels. This provides a unique synergy for both the ESCO and the equipment provider. Such a trade ally approach offers benefits in a large and scalable way.

Further advantages are offered by proper vendor-ESCO teaming, including:

d) Early project collaboration allows both the ESCO and vendor to compare notes on other items that can be brought to bear for a better and more complete solution.

An excellent example of project collaboration occurred when Dialight, Inc. and BGA, Inc. were comparing notes in their respective industries. Using the ESCO project approach, it became clear through these discussions that traffic signal projects can be enhanced beyond just the reduced electric demand and consumption.

Dialight, Inc. had been meeting with a municipality that was trying to budget an emergency battery back up system for its critical intersections. This municipality had been hit by a couple of hurricanes over the last few years with loss of power to those intersections. It became evident in Dialight's discussions with this municipality that because LED traffic signals use only 10% of the power draw of the existing incandescent lamps, a smaller battery backup system for its critical intersections would be required. This paradigm is not really surprising, as LED traffic signals represent one of the most efficient technologies available today.

This recognition offered an extreme project paradigm improvement. The synergy realized from combining the energy cost savings and maintenance cost savings, inherent with a typical LED traffic signal energy service project, with the reduced back-up installation cost was significant. The simple payback improved by 25%.

What was a compelling normal retrofit project became a very compelling project for the municipality to obtain a more cost effective traffic signal system and a much-needed battery back up system for critical intersections.

e) An extended warranty period provided by the equipment provider could enhance the savings value, and as such, increase the project value, which directly translates to project fee. An enhanced project fee realized by an extended warranty may very well more than justify the risk pricing of the extended warranty.

All parties win. The ESCO sells a larger project, the owner gets an extended warranty plus a larger project, and the equipment provider gets fair value for both the base product sale, and a fair return for the extended product warranty. Such opportunities for project optimization are best analyzed through a teaming arrangement.

f) An equipment provider is driven to continually update and improve its product line. An ESCO is driven to exploit every advantage an improved product provides, a) enhanced efficiency, b) longer life, and/or c) improved performance. Working together, the equipment provider gets full feedback more promptly while the ESCO gets to shape and fully exploit new and improved products.

Concurrently, the ESCO clients are early adopters of technologies. This is an inherent trait associated with considering innovative procurement means as demonstrated through procurement of energy services. Unfortunately, ESCOs, which must guarantee results, must avoid unproved technologies. With the vendor guaranteeing its product, the ESCO can offer a more cutting edge package to its client without a significant increase in risk.

g) A performance contract is all about managing risks. The typical risks associated with a performance contract are:

1. Project development costs
2. Project margin and product margin
3. Achieving the savings, M&V performance and guarantee obligation
4. Product warranty
5. Construction schedule
6. Payment risk
7. Project cancellation
8. Bonding requirements
9. Construction warranty
10. Sub-contractor performance

Sharing the Risks

A strong vendor-ESCO team could jointly address at least the first seven project risks.

1. **Project development costs**—Instead of the ESCO trying to sell the owner, and the vendor selling the ESCO and the owner, it could be much quicker and more cost-effective for the ESCO and vendor team to sell the owner, together. This is particularly helpful when the development is at risk.

2. **Project and product margin**—The best way to preserve both project and product margins is for the vendor and ESCO to work together to establish the most savings that can then be monetized into the project fee.

3. **Achieved savings, M&V performance and guarantee obligation**—A major risk taken by an ESCO involves project performance, which relies heavily on equipment performance. So, if the equipment provider already assumes the risk of equipment performance; i.e., product efficiency, the ESCO does not need to layer a equipment efficiency contingency. With equipment performance mitigated for the ESCO, less of the calculated savings needs to be set in performance reserve. The performance reserve may then be better used to pay the equipment provider for the performance warranty or to enhance the project.

4. **Product warranty**—As with product efficiency, the ESCO is taking a risk for product warranty. Additionally, the longer the warranty, the more savings are available to monetize to project fee. Working together, risk layering can be avoided and the optimum warranty term can be developed.

5. **Construction schedule**—Construction schedule is important because failure to meet this schedule can cause liquidated damages and construction financing charges. By working together, the vendor can anticipate product delivery schedules well in advance. Additionally, the vendor can coordinate drop shipping to the locations when the product is needed, which in turn minimizes product handling burdens. An involved vendor is more apt to give the ESCO's product delivery preference when production gets backed up.

6. **Payment risks**—Barring any performance issues, payment issues usually stem from the working dynamics between a client and a ser-

vice provider. If there are two companies, as suggested here, aligned with the project, there are two companies, which are able to manage the client relationship for payment.

7. **Project cancellation**—While it is rare for a PC contract to be cancelled after a delivery work order is let, it is still somewhat common for a client to cancel a PC project while in the development stage. This is where a vendor/ESCO relationship can really help each other, as both companies are aligned in keeping the owner interested in both the savings and the product.

Choosing The Correct Vendor-ESCO Partner

Obviously, for the vendor-ESCO team to be most synergistic, the ESCO should look for vendors that meet critical criteria. Summarized below are items that should be considered.

- The technologies selected should have high project-specific applicability
- Since the ESCO is looking for the vendor to step up to a higher-value role in the energy service project, it is critical for the right vendor to be selected. Critical to this selection is the financial strength of the vendor. Does the vendor have the balance sheet to stand behind its proffered risk mitigation elements of product efficiency and product life warranty?
- The vendor's reputation, experience and tenure in its industry. Since the ESCO is looking to the vendor to be part of its front-line team, even in the ESCO selection process, it is critical the vendor is not only known by the industry and the owner, but indeed enjoys a strong and favorable reputation.
- The vendor should be viewed as a value provider, not just a lowest first cost provider. Value includes much more than price; and
- The capacity of the vendor to deliver on its obligation needs consideration. Does the vendor have the distribution network to handle the product delivery needs?

For the vendor, key criteria in ESCO selection include:

- The reputation of the ESCO, and its success in winning project selections, should be strong.
- The track record of the ESCO's performance guarantees, to validate its skills in risk management, should be proven.
- The ESCO and equipment provider should have a compatible client base.

CONCLUSION

While the ESCO industry is all about providing best long-term value to an owner, because of the challenges of the ESCO industry, the industry itself does not always practice that philosophy in its project development and delivery process.

An attractive alternative to the normal ESCO project process involves early teaming with key vendors that represent cornerstone technologies to be used in the project. Benefits include alignment of risks, more cost-efficient development and delivery, a higher-value savings stream to monetize, protection of project and product gross profit margins.

Ultimately, it is the owner who directly benefits from the vendor/ESCO relation, which, at the end of the day, was the objective of the owner hiring the ESCO in the first place.

Reference

Hopper, Nicole, et al. Public and Institutional Markets for ESCO Services: Comparing Programs, Practices and Performances. March 2005. <http://eetd.lbl.gov/ea/EMS/EMS_pubs.html>

Chapter 15

Using the Web for Energy Data Acquisition and Analysis

Advances in new equipment, new processes and new technology are often the driving forces in improvements in energy management, energy efficiency and energy measurement procedures. Of all recent developments affecting energy management, the most powerful new technology to come into use in the last several years has been information technology—or IT. The combination of cheap, high-performance microcomputers together with the emergence of high-capacity communication lines, networks and the Internet has produced explosive growth in IT and its application throughout our economy. Energy information systems have been no exception. IT and Internet based systems are the wave of the future.

Timely energy information is particularly critical to energy service companies (ESCO's) as this information can be invaluable in establishing baseline energy consumption, commissioning new installations and ongoing monitoring of savings. Web-based energy information systems provide the ability to view time-stamped data on a daily basis, which provides both the contractor and the building owner the information to compare actual performance of the energy efficiency and/or conservation measures (ECMs) to the projected performance and to make adjustments to correct problems quickly. All of this can be done with just a web browser from the contractor's office, eliminating the need for costly site visits.

This chapter describes the fundamentals of an Energy Information System (EIS) and presents two case studies that showcase how energy data collection and analysis can be used to quantify energy project results.

Energy Information Systems

The philosophy, "If you can measure it, you can manage it," is critical to a sustainable energy management program. Continuous feedback on utility performance is the backbone of an Energy Information System[1].

A basic definition of an Energy Information System is:

> *Energy Information System (EIS):* Equipment and computer programs that let users measure, monitor and quantify energy usage of their facilities and help identify energy conservation opportunities.

Everyone has witnessed the growth and development of the Internet—the largest computer communications network in the world. Using a web browser, one can access data around the world with a click of a mouse. An EIS should take full advantage of these new tools.

EIS PROCESS

There are two main parts to an EIS: (1) data collection and (2) web publishing. Figure 15-1 shows these two processes in a flow chart format.

Data Collection Process

The first task in establishing an EIS is to determine the best sources of the energy data. Utility meters monitored by an energy management system or other dedicated utility-monitoring systems are a good source. The metering equipment collects the raw utility data for electric, chilled & hot water, domestic water, natural gas and compressed air. The utility meters communicate to local data storage devices by pre-processed pulse outputs, 0-10V or 4-20ma analog connections, or by digital, network-based protocols.

Data gathered from all of the local data storage devices at a predefined interval (usually on a daily basis) are stored on a server in a relational database (the "data warehouse"). Examples of relational databases are FoxPro, SQL and Oracle*.

Using an EMS for Data Collection

Identifying and organizing the best energy data sources is the first step in establishing an EIS. One potential data collection source is from an Energy Management System (EMS). An EMS typically has a built-in pro-

---

*Any reference to specific products or name brands of equipment, software or systems in this chapter are for illustrative purposes and does not necessarily constitute an endorsement implicitly or explicitly by the authors of this chapter or the others in this book.

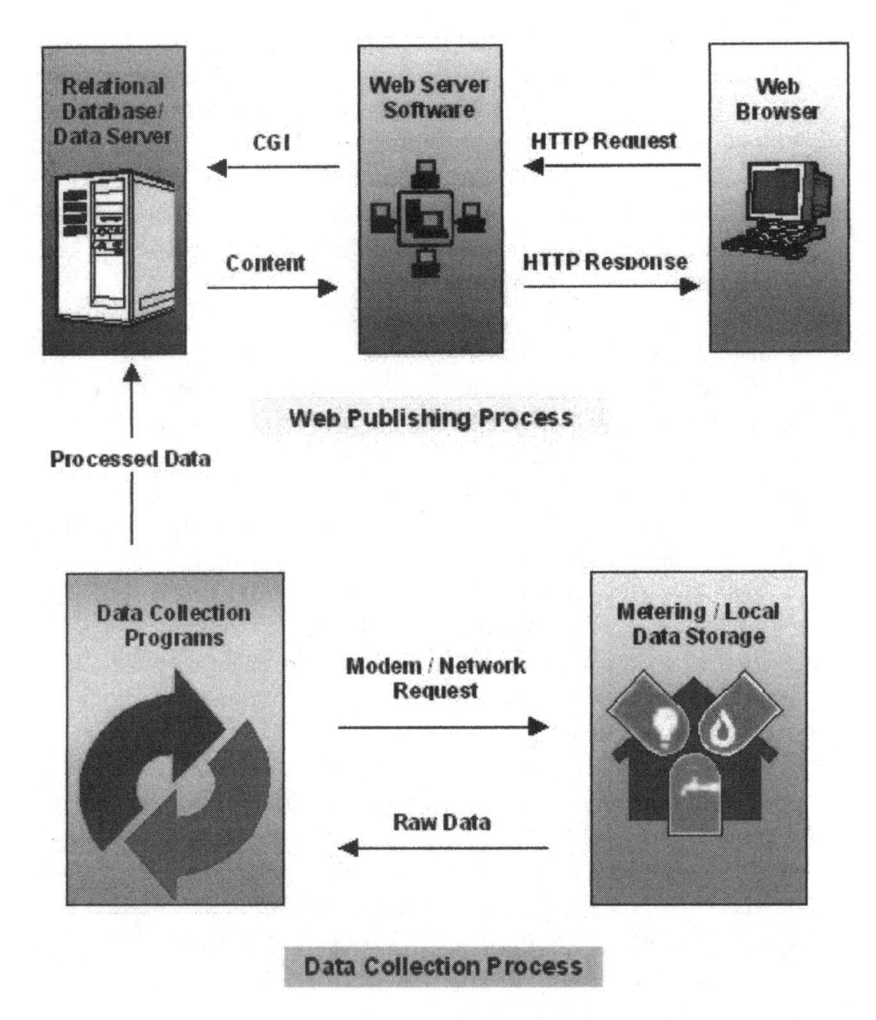

Figure 15-1. Energy Information System Schematic

cedure that can produce daily reports on points connected to the system. The process is to program the EMS to collect the desired utility data and then to move these data into the EIS relational database. Shown are the steps required for this to happen:

1. The EMS PC should be on the corporate LAN.

   This might be as simple as installing an Ethernet card in the existing EMS PC workstation. However, IT departments are generally very particular about the PC hardware installed on the corporate LAN and will probably require that the existing EMS PC be replaced with one of their standard PC's. This could cause compatibility issues with older EMS software operating on a new PC's operating system (OS) and require the EMS vendor to upgrade their software to a version that is compatible with the newer PC OS. So a seemingly simple task of putting the existing EMS PC on the corporate LAN might be turn out to be an expensive proposition.

2. Transfer the EMS Reports to the EIS Server on a daily basis.

   Once a path is established to the EMS PC front-end, the EMS report files need to transferred to the EIS Server. There are numerous methods to accomplish this task. This might be as simple as mapping a drive to the EMS PC from the EIS server. A DOS-based batch file can be launched at a specific time of day to copy the EMS report files to a subdirectory on the server.

   Another method would be to use File Transfer Protocol (FTP) to transfer files from the EMS PC to the EIS Server. Typically, a program on the EIS Server is required to coordinate the FTP file transfers.

3. Capture the data from the EMS Reports.

   Once the EMS reports have been moved to the server, a custom program is developed to extract the data from the EMS reports and update the EIS relational database. Ideally, this program is developed to read multiple EMS Vendors EMS reports. Since the relational database would have a standard format, the data from different EMS Vendors reports can be reported in a consistent format.

Collecting Data Using a DAS

Another approach to collecting utility data is to use a dedicated data acquisition server (DAS). The DAS allows users to collect utility data from existing and new meters and sensors. On a daily basis the DAS uploads the stored data to the EIS Server. Once the data have been transferred to the EIS Server, a program reads the DAS data files and updates the data in the EIS relational database for use by the web publishing program.

The AcquiSuite system from Obvius is typical of the emerging solutions and is a Linux-based web server, which provides three basic functions:

- Communications with existing meters and sensors to allow for data collection on user-selected intervals;
- Non-volatile storage of collected information for several weeks; and
- Communication with external server(s) via phone or Internet to allow conversion of raw data into graphical information.

The backbone of the system is a specially designed web server. The DAS provides connectivity to new and existing devices either via the on-board analog and digital inputs or the RS 485 port using a Modbus protocol. The analog inputs permit connection to industry standard sensors for temperature, humidity, pressure, etc., and the digital inputs provide the ability to connect utility meters with pulse outputs. The serial port communicates with Modbus RTU devices such as electrical meters from Veris, Square D and Power Measurement Ltd.

Web Publishing

The Internet, with the World Wide Web—or Web—has become quickly and easily accessible to all. It has allowed the development of many new opportunities for facility managers to quickly and effectively control and manage their operations. There is no doubt that web-based systems are the wave of the future. The EIS web publishing programs should take full advantage of these web-based technologies.

To publish energy data on the Internet or an Intranet (a private network that acts like the Internet but is only accessible by the organization members or employees), client/server programming is used. The energy data are stored on the EIS server, and wait passively until a user, the client, makes a request for information using a web browser. A web-publishing program retrieves the information from the EIS relational database, and sends it to the web server, which then sends it to the client's web-browser which requested the information.

There are many software choices available for the web-publishing process. One method uses a server-side Common Gateway Interface (CGI) program to coordinate the activity between the web-server and the web-publishing program. CGI is a method used to run conventional programs through a web browser.

The web-publishing client/server process for an EIS uses the steps below (See Figure 15-1). This entire process takes only seconds depending on the connection speed of the client's computer to the web.

1. A user requests energy information by using their web browser (client) to send an HTTP (Hypertext Transfer Protocol) request to the web server.

2. The web server activates the CGI program. The CGI program then starts up the web-publishing program.

3. The web-publishing program retrieves the information from the relational database, formats the data in HTML (Hypertext Markup Language) and returns it to the CGI program.

4. The CGI program sends the data as HTML to the web server, which sends the HTML to the web browser requesting the information.

Web Publishing Programming Options

There are many programming alternatives available other than the CGI approach described above. PERL, Active Server Pages (ASP), JavaScript, and VBScript, Java Applets, Java Server Pages, Java Servlets, ActiveX controls and PHP are a few of the more popular choices available today. Some of these are easier to implement than others. ASP for instance, is a part of IIS, so no installation is required. PERL and PHP require installation of their respective programs on the web server machine to run. There are also security issues with some of these approaches. The client's machine downloads Java Applets then executes them from there. Some experts view this as a security risk not worth taking. Javascript and VBScript are somewhat limited in that they are just a subset of the other full fledged programming languages. Most browsers interpret them correctly so no installation is required. Java Server Pages and Java Servlets run on the web server in the same way as ASP, but may require some installation depending on the web server used.

Although there are many web servers available to choose from, two are the most popular by far are Microsoft's Internet Information Services (IIS), which comes with Windows 2003 server, and Apache web server, which is a good choice for other operating systems. Any web server needs some configuration to produce web content, especially if it is querying a database. The web-publishing task will likely require custom folders, special access permissions and a default page.

After installing the web server, the web-publishing administrator must put a default page in the root directory of the web server. This is the first page users will see in their browser when they type in the web site's Internet address. The pages are usually named "default.htm" or "index.htm" but can be anything as long as the web server is configured to treat them as the default page. Next, if CGI is used, the administrator creates a special folder to store the scripts. This is usually called "cgi-bin" or just "scripts." This folder must have permissions specifically allowing the files in the folder to be "executable." In some cases, "write" permissions are required for the folder if the CGI programs write temporary files to it. Other custom folders may be required to organize the web publishing content. Once the web-publishing administrator configures the web server, he or she can install and test custom CGI programs and pages. If the CGI program or pages accurately return data from the database, then the task of creating custom reports for the energy data can begin.

EIS Implementation Options

Deciding which web server and programming method to use along with configuring and implementing it to create a web publishing system can be quite a task. It really requires an expert in these areas to do a reliable job. Three approaches have evolved to satisfy web-publishing requirements.

1. Use internal resources to accomplish this task. This works well if there are already experienced web programmers available and they have time to work on the project. This makes it easy to customize the web publishing content as needed quickly and cost effectively. Finding time for internal personnel to focus on the project is usually the problem with this option.

2. Hire an outside consultant to do the configuration and programming as needed. This works well if the consultant has a good working relationship with someone internally to facilitate access to the protected systems and help with understanding the data. The consultant must be willing to work for a reasonable rate for this approach to be cost effective. The consultant must also be responsive to requests for support.

3. Purchase and install a somewhat "canned" version of the web publishing software and then customize it to fit the energy data as required. This approach has many possible problems in that the software is usually quite expensive and requires a great deal of customization and support from the outside to make it work well. However, for small simple projects this may be a good fit.

For users, who do not want to invest the time and effort required for this "do-it-yourself" approach, there are numerous companies that provide a complete EIS service for an on-going monthly service fee. The EIS service company provides all of the IT-related functions, including the energy data collection/storage and the web-publishing program. The user accesses the EIS service web site using a web browser, enters a User ID and password and then uses the available reports/graphs to analyze energy data.

The advantages of this approach is that the user does not get involved with the details and operation of the EIS, but instead is able to work with the EIS service provider to develop the utility data reports most helpful to their operation. The downside to this approach is the on-going monthly service fee that is a function of the amount of data processed—the more meters or bills processed the higher the monthly fee. There may also be additional costs to customize any reporting from the standard reports already created by the EIS service provider. The Building Manager Online service from Obvius* is one of the many choices available to users today.

MEASUREMENT AND VERIFICATION†

The majority of energy saving retrofit projects are implemented based on engineering calculations of the projected return on investment [2]. As with any projections of ROI, much of what goes into these calculations are assumptions and estimates that ultimately form the basis for implementation. As the folks at IBM used to say, "garbage in—garbage out," which in the case of energy retrofits means that if any of the assumptions about parameters (run times, set-points, etc.) are wrong, the expected payback can

*Jim Lewis is one of the authors of this chapter and is the CEO of Obvius, LLC.
†See also Chapter 4 for a more detailed discussion of measurement and verification and accepted protocols.

be dramatically in error. The establishment of good baselines (measures of current operations) is the best way to determine the actual payback from investments in energy and sub-metering.

Just as important as building an accurate picture of the current operation is measuring the actual savings realized from an investment. If there is no effective means of isolating the energy used by the modified systems, it may be impossible to determine the value of the investment made. Using monthly utility bills for this analysis is problematic at best since the actual savings achieved can be masked by excessive consumption in non-modified systems.

Consider, for example, a commercial office building which has a central chiller plant with an aging mechanical and control structure that provides limited capability for adjusting chilled water temperature. To improve efficiency, the building owner plans to retrofit the system to provide variable speed drives on pumps for the chilled water and condenser water systems along with control upgrades to allow for chilled water set-point changes based on building loads. In the absence of baseline information, all calculations for savings are based on "snap-shots" of the system operation and require a variety of assumptions. Once the retrofit is completed, the same process of gathering snapshot data is repeated and hopefully the savings projected are actually realized. If the building tenants either add loads or increase operational hours, it is difficult if not impossible to use utility bills to evaluate the actual savings.

In contrast, the same project could be evaluated with a high degree of accuracy by installing cost-effective monitoring equipment prior to the retrofit to establish a baseline and measure the actual savings. While each installation is necessarily unique, building a good monitoring system would typically require:

- Data acquisition server (DAS) such as the AcquiSuite from Obvius to collect the data, store it and communicate it to a remote file server.

- Electric submeter(s)—the number of meters would vary depending on the electric wiring configuration, but could be as simple as a single submeter; e.g., Enercept meter from Veris Industries, installed on the primary feeds to the chiller plant. If desired, the individual feeds to the cooling tower, compressors, chilled water pumps, etc. could be monitored to provide an even better picture of system performance and payback.

- Temperature Sensors (optional)—in most installations, this could be accomplished by the installation of two sensors, one for chilled water supply temperature and the other for chilled water return temperature. These sensors do not provide measurement of energy usage, but instead are primarily designed to provide feedback on system performance and efficiency.

- Flow Meter (optional)—a new or existing meter can be used to measure the gallons per minute (gpm). By measuring both the chiller input (kW) and the chiller output (tons), the chiller efficiency can be calculated in kW / ton.

The benefits of a system for actually measuring the savings from a retrofit project (as opposed to calculated or stipulated savings) are many:

- The establishment of a baseline over a period of time (as opposed to "snapshots") provides a far more accurate picture of system operation over time.

- Once the baseline is established, ongoing measurement can provide a highly accurate picture of the savings under a variety of conditions and establish a basis for calculating the return on investment (ROI) regardless of other ancillary operations in the building.

- The presence of monitoring equipment not only provides a better picture of ROI, but also provides ongoing feedback on the system operation and will provide for greater savings as efficiency can be fine-tuned.

Viewing and Using the Data

Historically, much of the expense of gathering and using sub-metering data has been in the hardware and software required and the ongoing cost of labor to produce useful reports. Many companies are leveraging existing technologies and systems to dramatically reduce the cost of gathering, displaying and analyzing data from commercial and industrial buildings. The AcquiSuite data acquisition server uses a combination of application specific hardware and software. A standard web browser, such as Microsoft Internet Explorer, provides the user interface.

The AcquiSuite DAS automatically recognizes devices such as meters from Power Measurement Ltd. and Veris Industries, which makes installation cost effective. The installer simply plugs the meters in the DAS and all configuration and setup is done automatically with the only input required being the name of the device and the location of the remote file server. The DAS gathers data from the meters on user-selected intervals; e.g., 15 minutes, and transmits it via phone line or LAN connection to a remote file server where it is stored in a database for access via the Internet.

To view the data from one or more buildings, the user simply logs onto a web page; e.g. *www.obvius.com*, and selects the data to view.

The gathering and sorting of data do not provide sufficient energy management guidance unless the data are analyzed, transformed into usable information, and implemented. To help illustrate this point, the following case studies are offered.

CASE STUDY—RETAIL STORE LIGHTING

BACKGROUND

A retail store chain in the Northeast was approached by an energy services company about converting some of their lighting circuits to a more efficient design. On paper, the retrofit looked very attractive and the company elected to do a pilot project on one store with a goal to implementing the change throughout the entire chain if it proved successful. The retailer decided to implement a measurement and verification (M&V) program to measure the actual savings generated by comparing the usage before the retrofit and after.

IMPLEMENTATION

The store had 12 very similar lighting circuits, all of which were operated on a time schedule from a central control panel in the store. Since the circuits were very similar, it was decided that measuring the impact on one circuit would provide a good indication of the savings from the other circuits. The M&V equipment consisted of the following:

- An electrical sub-meter (see Figure 15-2) was installed on the power lines feeding the lighting circuit;

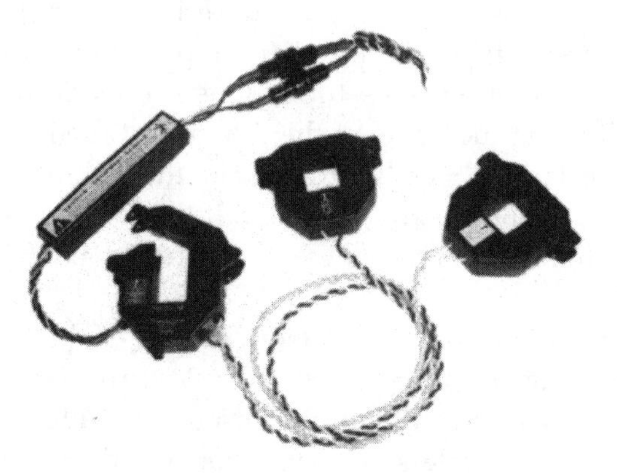

Figure 15-2. Retrofit Electric Sub-meter

- A data acquisition server (see Figure 15-3) was installed in the store to record, store and upload time-stamped interval data to a remote server for storage and display. The DAS provided plug and play connectivity to the sub-meter and used an existing phone line or LAN to send data from the store to a remote server on a daily basis.

Figure 15-3. Data Acquisition Server

- The remote server was used to monitor consumption before the retrofit and to measure the actual savings

RESULTS

Figure 15-4 shows the actual kW usage over roughly 24 days. The left side of the chart shows the kW usage for the first 11 days before the retrofit and the average usage is fairly constant at around 1.45 kW. On Feb. 11, the retrofit was performed, as indicated by the drop to zero kW in the center of the chart. Immediately after the retrofit (the period from Feb. 11 to Feb. 15, the kW load dropped to around 0.4 kW, a reduction of over 70% from the baseline load in the left of the graph.

The good news for the retailer was that the retrofit performed exactly as expected and the M&V information obtained from monitoring the energy on this circuit provided clear evidence that the paybacks were excellent. The initial good news, however, was tempered somewhat after looking at the chart. It was immediately evident that this lighting circuit (and the other 11 identical circuits) were operating 24 hours per day, seven days a week. The store, however, operated from 10 AM to 9 PM each day and the lighting panel was supposed to be shutting off the circuits during non-operating hours.

The electrical contractor was called in to look at the system and determined that a contactor in the panel had burned out resulting in continuous operation of the lighting circuits throughout the store. Once the contactor was replaced, the operation of the lighting panel was restored so that the lights were only on during operating hours and shut off during the night, as indicated by the right side of the chart.

This simple chart of energy usage provides an excellent example of two uses of energy information:

1. **Measurement and verification of energy savings**—The left hand side of the chart clearly shows the actual energy reduction from the lighting retrofit and the data provided can be used to extrapolate the payback if this same retrofit is applied throughout the chain; and

2. **Use of energy information to fine-tune building operations**—In addition to the M&V benefits of energy information, this example also shows how a very simple review of energy usage can be used to make sure that building systems are operating properly.

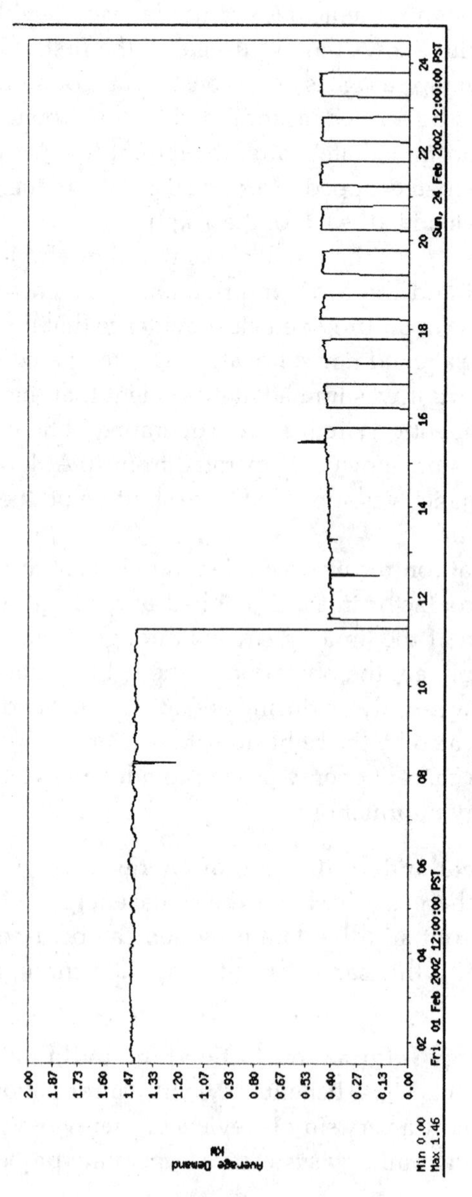

Figure 15-4. KW Loads for a 24-Day Period

CASE STUDY— CHILLER PLANT OPTIMIZATION

Chiller plants consume an enormous amount of energy to produce the air conditioning required for a building. By sub-metering the chiller operation, the operator can directly measure the effects of changing set points and time schedules on energy usage.

BACKGROUND

This case study is the result of the optimization effort at a new chiller plant built for a large convention center. The mechanical engineer designed the chiller plant as a variable flow primary system. An energy management system controlled the operation of each of these chiller plant operations.

Sub-metering of the chiller plant operation was part of the original design for the convention center. Each chiller came equipped with an on-board electric meter, chilled water supply and return temperature sensors. A chilled water flow meter was added to each chiller to allow for the chilled water tons to be calculated. Additional electric sub-meters were included to measure the condenser and chilled water pump motors and the cooling tower fans. The energy management system collected the energy data, chiller operational parameters and the convention center space temperature and relative humidity values.

Each night, the energy management system produced a report file for the data being trended. This file was stored on the energy management system server. A custom program read this file and reformatted the data and added this to a relational database along with all the other data collected from other projects. This is the data collection process shown in Figure 15-1 previously.

A custom program was developed to display and graph the energy data on a user's PC using a web-browser. Another custom program pushed the energy data report via email to the convention center maintenance personnel. These reports provided the information to the user so they could determine if the chiller plant was operating efficiently. The process to pull the data from the relational database and produce reports was the web-publishing process shown in Figure 15-1 on page 245.

CHILLER PLANT OPTIMIZATION

As with most new projects, there is a period of test and adjustment that results when going from the drawing board into actual operation. In consultation with the design engineer and the chiller manufacturer, several changes resulted in reducing the chiller plant energy usage by approximately 30 percent.

The chiller plant had several energy saving features, which are used to control the operation. The main control parameters are shown below:

- Each chiller had a leaving water set point and operational schedule.

- Chilled water pumps maintained a differential pressure at the farthest air handler unit. Variable speed drives were used to vary the motor speed and the resulting pump flow.

- A chilled water bypass valve is operated to maintain a minimum flow rate through the chillers when the chiller plant flow to the building was low.

- The condenser water pumps were interlocked with the chiller operation.

- Cooling tower fans are controlled by variable speed drives to maintain a condenser water supply temperature set point back to the chillers.

- The air handlers in the convention center exhibit space were set up as single-zone variable air volume systems. The discharge air temperature was fixed and variable speed drives modulate the fan speed to maintain the space temperature set point.

- CO_2 sensors were used to measure the occupancy. If the CO_2 sensor was below the set point, the outside air dampers were kept at their minimum values. As the CO_2 levels increased, the outside air dampers were modulated open to maintain the CO_2 levels at set point.

The first operational issue was the result of the chillers tripping offline from surging. After studying the problem, it was determined that the humidity control programmed for the air handler chilled water valves caused rapid flow rate changes resulting from the chilled water valves opening and closing to maintain humidity set points. The solution was to change the air handler chilled water control algorithms to maintain a fixed discharge air temperature. Fortunately, each air handler was equipped with a variable speed drive. The control algorithm was changed to ramp the supply fan up and down to maintain the space temperature. The humidity in the convention space actually improved using this control strategy.

Once the chiller operational issues were solved, the next focus was to reduce the energy usage in the facility. The first changes resulted from lowering the minimum flow rate set point for each chiller from 600 gpm to 400 gpm. This resulted in an increased differential temperature at each chiller. The next change was to increase the chilled water leaving temperature to 44 degrees F from the original design value of 40 degrees F. Both of these adjustments combined to result in a 19% energy reduction at the plant.

The next operational change was to shut the entire chiller plant down from midnight to 6 a.m. Each air handling unit and all exhaust fans were also simultaneously shut down. The temperature and humidity values in the convention center were trended and did not show any significant increases. They quickly recovered in the morning before any convention activity occurred. This change in operation resulted in a further reduction of 15 percent in the chiller plant energy usage.

An operational problem was detected when the chiller plant daily report data showed a significant increase in the chiller plant electric usage even though there was no convention activity. It turned out that the chiller maintenance company had manually forced the chillers for testing purposes, but failed to put them back into automatic control causing one chiller to run from midnight to 6 a.m. The problem was quickly resolved and the chiller plant returned to automatic operation.

CHILLER PLANT DATA ANALYSIS TIMELINE

Figure 15-5 shows the raw data and timeline of events for the chiller plant optimization. The data were measured on a hourly basis and summarized into daily totals. The web-based energy information system captured all of the data and was updated daily. The EIS allowed the users to inter-

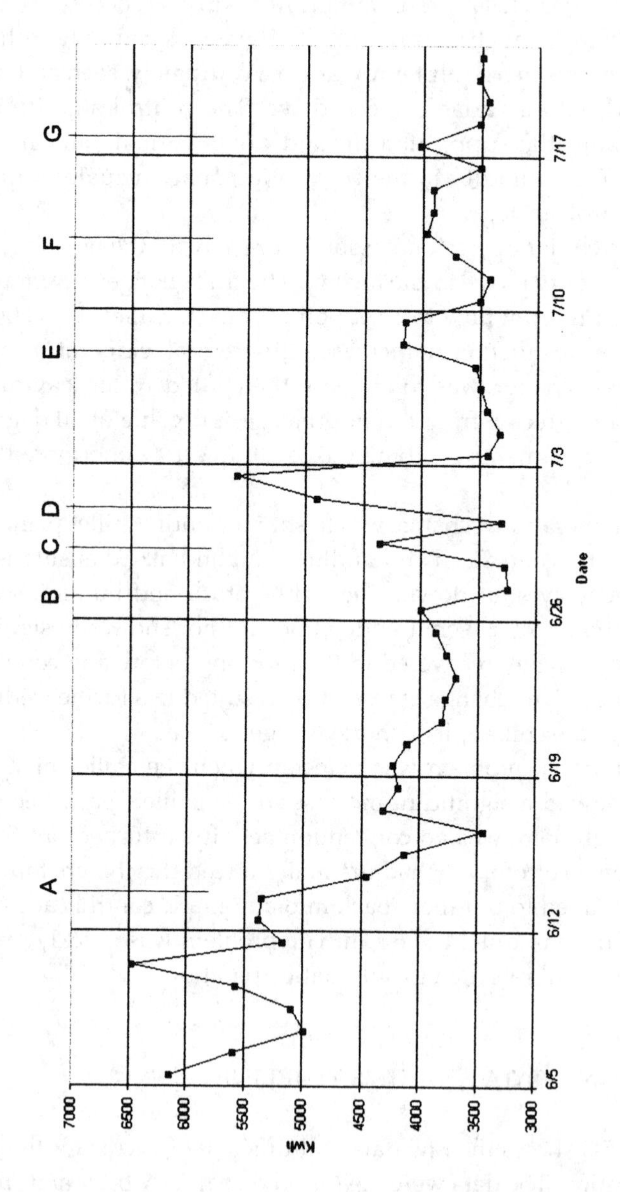

Figure 15-5. Chiller Plant kWh Usage

rogate the data and measure the impact of operational changes. In making comparisons to determine operational changes, the daily totals could be matched against ambient weather conditions and convention center activities to ensure an apples-to-apples comparison. By also trending the convention temperature and humidity values, operational impacts resulting from the energy conservation strategies could be quantified as well.

A. On June 14th, the chilled water set point was raised from 40F to 44F. Additionally, the minimum flow through the chillers was lowered from 600 gpm to 420 gpm.

Event	Outside Temp	Avg kWh	kWh Change	% Change
Before Reset	82.3F	5,192		
After Reset	83.0F	4,205	-987	-19%

B. Starting on June 27th, the chiller plant was completely turned off from midnight to 6 p.m.

Event	Outside Temp	Avg kWh	kWh Change	% Change
Before Shutdown	80.7F	3,873		
After Shutdown	81.1F	3,271	-602	-15%

C. On June 29th, a Test and Balance company performed some performance tests on the chiller plant that resulted in excessive chiller plant operation.

D. On July 1st and 2nd, the convention center was occupied with a very large event increasing chiller plant demand significantly.

E. On July 8th, the chiller maintenance company mistakenly forced the chillers out of automatic control which caused a chiller to operate from midnight to 6 p.m. The mistake was corrected on July 10th.

F. On July 13th, the convention center was occupied with convention activity.

G. On July 17th, the convention center was occupied with convention activity.

By using the electrical submetering data as feedback, adjustments to the chiller plant control strategies were made that resulted in a 30 percent reduction in energy usage from the original sequence of operation.

CONCLUSION

Historically, hardware, software and installation of energy information systems have been prohibitively expensive and has limited implementation to those commercial and industrial facilities that could afford to pay for custom systems integration services. These costs have fallen dramatically as companies leverage the enormous investment in the Internet to provide tools to the building owner that make do-it-yourself data acquisition a cost effective reality. Hardware and software designed specifically for data acquisition, and using available tools such as TCP/IP, HTTP and Modbus, puts valuable energy information literally at the fingertips of today's facility owners and provide an excellent method for measurement and verification of energy saving projects.

Successful ESCO's employ the latest technologies to help their customers reduce energy consumption and lower operating costs. Web-based Energy Information Systems provide one more valuable tool for the contractor and the building owner in managing energy by providing timely and accurate measurement of performance for individual systems or total buildings.

Bibliography

[1] Barney Capehart, Paul Allen, Klaus Pawlik, David Green, *How a Web-based Energy Information System Works,* Information Technology for Energy Managers, The Fairmont Press, Inc., 2004

[2] Jim Lewis, *The Case for Energy Information,* Information Technology for Energy Managers, The Fairmont Press, Inc., 2004

Section IV

ESCOs Go Global

When ESCO pioneer, Don Smith, was asked why he wanted to expand his market overseas, he responded, "That's a no brainer. That's where the market is."

Today more than ever: "That's where the market is." A market with huge opportunities—and lots of problems. The energy efficiency market in non-OCED countries was estimated near $20 Billion a decade ago. Since then, there has been considerable growth in non-OCED as well as industrialized countries, and the opportunities are even greater.

Once ESCO management starts thinking about foreign markets, there is a moment of euphoria when it feels like the world is its playground. The temptation is great to jump on a plane and go see what's out there. At first blush, it seems like one big wonderful opportunity. But what does it take for an ESCO to export its expertise to a country that may not have the business, legal or financial infrastructure to support an effort in a yet-to-be-created market?

Even before an ESCO starts to consider these factors, the firm must first assess its own capabilities and its readiness to move overseas. ESCOs will find even greater risks in foreign markets and these risks will become more varied as an attempt is made to use unfamiliar risk mitigation strategies.

This section is devoted to taking a hard look at the opportunities and the difficulties associated with the international market place. It starts with some tough questions and important answers posed by Don Smith. Don did such an excellent job of analyzing an ESCO's internal situation in the first edition, it has been included here verbatim except for the title. Don has retired from the ESCO industry, but we are still indebted to him for the excellent leadership he gave us when he headed up Viron and later Energy Masters.

Once an ESCO has decided it has what it takes to go overseas, then it must decide which country to target and what market segments to focus upon within that country. Drawing on experiences in ESCO devel-

opment in 34 countries and a proven international assessment protocol, which Kiona developed, Jim Hansen offers must-read guidance in chapter 17 designed to help ESCOs weigh global options. Jim is president of Kiona International, which specializes in ESCO development around the world.

Whenever performance contractors with overseas interests gather, the topic inevitably turns to financing. The lack of commercial financing is a pervasive obstacle for ESCO development around the world. We are indeed fortunate to have the "guru" of international energy efficiency financing, Tom Dreessen, offer us his vision of the direction energy efficiency financing needs to go if local commercial banks will be there to support the growth of ESCOs in a given country. After years of heading up domestic and international ESCOs, Tom now focuses on the most critical need in energy efficiency work, financing. As President and CEO of EPS Capital Corporation, a former bank trustee, and a CPA, Tom gives us the benefit of his insights into the critical elements of cash flow financing in

Chapter 18.

This section is capped off by a glimpse of the global marketplace by Bob Dixon. Bob is the head of the global market for energy services for Siemens and immediate past president of the National Association of Energy Service Companies. From these experiences and his years in the ESCO industry, he presents an excellent synopsis of the competencies ESCOs must have to go abroad. While he states that there are exciting opportunities for ESCOs in the global economy, he stresses that it will only work if the ESCO has the cultural sensitivity, receptivity to new ideas, and the ability to work within the local resource acquisition situation.

Bob's observations for the global marketplace are equally applicable to the domestic scene. He underscores the thoughts of Don, Jim and Tom when he notes that the international ESCO business can be exciting, but it is not for the faint of heart. Bob has thrown down the gauntlet in a strong challenge to ESCOs who want to grow. It is a great way to conclude this second edition.

Chapter 16

Are You Ready to Go International?

To go international… or not to go? That is the question.

Paraphrasing the Bard doesn't make the decision to expand into the international market any easier, particularly for those in the complex energy-efficiency performance contracting business. Because energy services company (ESCO) projects carry certain risks, logic tells us an international setting would only magnify those risks.

True enough. But that is only *part* of the answer. The real question is: *Can you afford NOT to go international?*

Unlike Shakespeare's Hamlet, you HAVE to make a decision—especially in today's brutally competitive marketplace. To help with the decision-making process, I've broken it into the following general questions, which we'll explore in greater depth in this chapter.

1. *Is the international marketplace worth pursuing?*
2. *Is your company ready to go international? Can it handle an international venture without disrupting your domestic business?*
3. *Can you afford it financially?*
4. *Which countries will you target?*
5. *Should you have a local partner? What roles will you and your local partner play, both now and long term?*
6. *What value will your company bring to ensure a successful, long-term relationship and not just a short-term technology transfer?*

Q: The international market: Is it worth pursuing?

A: This is a no-brainer. For an energy services company to become a dominant player and maintain that position, the ESCO *must* participate in the international marketplace. Most recent studies show that the potential for ESCO services *outside* the United States *far ex-*

ceeds the U.S. potential, even though many countries need considerable development before an ESCO can become profitable there.

Q: ***Are you really ready to go international?***
A: To answer that question, you must first ask yourself this question: *Is your business in order in the United States?* Unless you answer "yes," you can't expect to add international business with any reasonable hope of success.

When my company decided to go international, we actually put international plans *on hold* for six months while we finished up some U.S. business first: achieving sales goals, making sure comprehensive business systems were in place, etc. Our reasoning? Developing an international presence takes so much time and energy that it's not feasible to catch up on the domestic front at the same time.

Q: ***Has your company successfully developed its management, risk control, technical and other systems for its U.S. business and replicated them in remote U.S. offices?***
A: Key to success in the complicated business of energy services is developing and fine-tuning your technical, business and management systems to operate with consistency. You must be able to evaluate and manage risk and follow similar procedures each time, using the same forms and software for technical audits and analyses.

In addition, your technical, management and marketing systems should tie into your business accounting software to ensure that the entire company runs smoothly, seamlessly and profitably.

Project managers should use one project tracking and management structure consistently, enabling them to see in an instant precisely how their jobs are doing so they know when to adjust. Executives, likewise, must have up-to-date information to keep the overall business on track and make course corrections at the appropriate time.

More importantly, you must be able to replicate these systems in remote offices so that all of your people have access to the same information and do essentially the same tasks in the same way. Until your company does this successfully throughout the United States, you won't want to attempt it elsewhere.

The international equation is further complicated by different business cultures, customs, laws and languages, plus the great geographical distance separating employees from their home base.

Q: ***Do you have in-house management expertise and time available to dedicate to international work?***

A: You must determine the management level and amount of management time needed for the task—and what you are willing to spare.

Further, ask yourself: *What tasks must be completed? What level of decision-making is required? How much management time will be spent abroad?*

Based on my experience, I can assure you that doubling or tripling your answers is prudent. Whatever your answers, can your company afford to lose key management time from your core business?

When we did this exercise for our first international venture, we determined we had to dedicate two man-months out of 12, or 1/6 of one key executive's time, the first year. Fortunately, we had international executive experience in house. In addition to this management commitment, we had to dedicate considerable technical and support staff to the effort.

Q: ***Along with management and human resource expenditures, can you afford the financial expenditure? How much will it cost?***

A: Influencing cost are such factors as the actual international location, the partner, the market and others. Somewhere early in the planning stage, you've got to get a handle on cost.

The following are just some of the costs you'll want to consider. Some are better handled by local partners; others, directly by your company.

- Local office space and support staff
- Sales literature and other marketing support. (If you think the sales lead time for a performance contract is long in the United States, think what it might be in another country, particularly one unexposed to the idea of performance contracting.)
- Sales expense
- Technical support for the sales staff
- Business plan development
- Travel and *per diem* costs

- Training personnel (your local partner's and your own)
- Adapting software programs to another country
- Project engineering and management
- General management and executive personnel

If you think your first international venture will make money or break even the first year, or even the second, *think again*. You should be prepared to fund the venture for a full year, and possibly two years or more, *with* LITTLE OR NO REVENUE *flowing back to your company.*

Q: ***Have you evaluated your target countries?***
A: Once you've made the decision to go international, you must target a specific country or part of the world where you want to do business. This raises a whole new set of questions.

Q: ***How big is the market for your services in the target country? Particularly, for the vertical markets that your company favors?***
A: The market must be large enough to make your venture worthwhile. One way to assess market size for energy-efficiency performance contracting is first to examine the total energy market, or the amount spent for electricity and fuel to operate, cool and heat buildings—and then make some assumptions.

For example, assume the country you're interested in has a total energy market of $1 billion. Your research further shows that half, or $500 million, is used for commercial buildings and the other half, for residential and industrial.

If you're interested solely in commercial, you can concentrate on the $500 million number. In the course of your research, you've also determined that few or no energy retrofits have been implemented in commercial facilities and it would be reasonable to achieve a 20 percent reduction in energy use by implementing cost-effective retrofits with three-year pay backs.

In this example, the potential total retrofit market for commercial facilities would be **$500 million x .2 x 3 = $300 million.**

The really big question is how much of this market could your company secure on an annual basis? A one percent (1%) penetration would yield $3 million; 5 percent would yield $15 million; 10 percent would yield $30 million, and so on.

The real answer, of course, lies with your sales and implementation ability, the extent of competition, local acceptance of performance contracting and a host of other variables.

Ultimately, you'll need good business planning and financial *pro formas* to determine if, and when, you can make a profit, and how much.

Q: ***Do you understand the laws and business customs of your target country?***

A: Certain issues are especially important, such as ease of repatriating money, taxes and other factors affecting an American company doing business in that locale.

Q: ***What are the political and economic climates like now? What does the future hold?***

A: Ideally, you'll choose a country with a stable political environment, a good economy and positive, controlled growth. You don't want to be enmeshed in wars, political insurrections, runaway inflation and the like.

Q: ***Is competitive project financing readily available?***

A: This is an extremely important issue and more than a little complex. You can't do projects unless readily accessible and competitive capital is available.

If you plan to deal primarily in the public sector, will special low-interest rates be available, such as tax-exempt rates in the U.S?

What are the prevailing rates in both the public and private sectors?

Will you use traditional third-party financing in which the customer signs a note directly with the financing institution, or are you looking at some form of shared savings?

Prevailing corporate returns also can determine whether a project is a "go." If companies can secure a 50 percent ROI by investing capital in their own core businesses, why would they want to invest in energy efficiency projects with a 30 percent ROI? In this scenario, the project sale may depend on off-balance sheet financing in which the customer truly is NOT investing any capital.

Q: ***Do American banks have a significant local presence?***

A: If American banks are already committed to investing heavily in another country, chances are better for having your performance contracting proposal embraced.

Can 100 percent project financing be secured, or will you or your customers have to make an equity investment in the project? Do the financial institutions in the host country understand the concept of performance contracting, and are they willing to advance funds for both construction and permanent financing? What kind of terms are available? Will they go 10 years?

The above questions give you a flavor of the complexities involved with international project financing, which really is a subject unto itself.

Q: ***Are the right factors in place to make performance contracting projects feasible?***

A: Here's a check list to review:

- Energy rates: high, low or middle-of-the-road?
- Are the buildings generally inefficient?
- Does the government back energy efficiency?
- Is financing available for energy-efficiency projects?
- What is the status of the utility industry? Is it fully regulated or deregulated? What about load factors? Demand side management programs?
- How good is the building stock infrastructure?
- Are top-notch professional services available?
- Are good subcontractors available? Are labor rates favorable?
- What are the vertical markets like?

In some countries, like India, the market is more than 80 percent industrial. If your company's expertise and target markets lie elsewhere, then India may not be the best choice for your international venture.

Q: ***How important is a local partner?***

A: A local partner is essential. Going international without a local partner would be worse than trying to do corporate taxes without an accountant or legal work without an attorney. There are too many things to learn, too many customer relationships to develop and too many unknowns for you to go it alone.

Before evaluating a potential local partner, you'll want to consider the roles you and your partner will play:

- Since sales are built around developing personal relationships, you'll want your partner to have a significant role in sales even though you might be using your own proven sales process. If this is your approach, the ideal is to find a partner with a good customer base and sales personnel.

- Since your own technical resources, such as project engineers and managers, technicians and the like are limited—and the project's success depends on technical creativity and doing things right—you might want a partner with strong local technical resources.

- Because these projects will be at a great distance geographically, with more variables and unknowns than you're accustomed to, you might want to share financial risks with your local partner. This means you'll want to choose a partner who is strong financially.

You'll also want the following from your local partner:
- positive reputation;
- good management team; and
- commitment to success.

Q: ***What are some of the important issues in structuring the business relationship?***

A: These issues are important to consider:
- You'll want good input from your outside accountants and financial advisors on whether to form a partnership, joint venture or contractor-subcontractor relationship.
- How will you share profits?
- How will you share risks? Sharing profits probably will be somewhat proportional to sharing risks.
- Who pays start-up costs? Are they later folded into job costs and reimbursed?
- What steps must you take to ensure that you'll add value to the relationship year after year rather than simply transferring technology to your partners, which they'll later utilize on their own?

Q: ***How does one begin all this research?***

A: You have lots of good resources: Your local public or university library, the U.S. Chamber of Commerce, banks, the National Association of Energy Services Companies (NAESCO), the U.S. Agency for International Development (AID) and other federal agencies; or perhaps you'll be able to hook up with a good agent to guide you.

The time to start investigating and analyzing is *now*. Others are moving forward already.

Chapter 17

Assessing International Opportunities

With the growing acceptance of globalization as fact rather than a controversial theory, more and more companies, smaller than the "Fortune 500," are looking beyond our shores for opportunities to do business. With the maturing of the ESCO industry in the United States, the advance of retail access, subsequent price volatility, rapid changes in the utility industry and the general move toward a "global economy," is it time for global ESCOs?

The size of the global market is huge and the efficiency possibilities are almost beyond imagination. However, the "global market" is not just one big plum waiting to be harvested, it is many, many individual and varied markets. The way to tap those markets may be different than anything you have seen before. Thinking "outside the box" is a necessity.

International markets can be very favorable targets for an ESCO that is ready to broaden its market base. ESCOs from a number of countries are already active around the world. British, German, Austrian, French, and Australian companies, and others, are looking for business. Some are quite firmly established far from their homes. A small number of American ESCOs have taken steps across borders, but the number and scope of these efforts has been quite limited.

Not long ago we spoke of "developing countries" such as the Central European nations, the new nations that once formed the Soviet Union, the "awakening giants" like China, Indonesia and India. Well, they are still developing, and they are a lot farther along today than they were just a few short years ago. Many countries have been exposed to the concept of performance contracting through the efforts of organizations such as the US Agency for International Development (USAID), the World Bank, Asian Development Bank (ADB), the European Bank for Reconstruction and Development (EBRD) and others. Opportunities for the energy service industry are still there. In fact even better opportunities exist, as the economy and level of business sophistication within these nations has grown. The markets in the wide world may no longer be "virgin territory" for an adventurous ESCO, but there are very real opportunities if international expansion is approached in a systematic, businesslike way.

PRELIMINARY ASSESSMENT

For the ESCO serious about venturing forth into the global marketplace, it will pay handsomely to first make an internal assessment of the firm's capabilities and resources as suggested by Don Smith in the previous chapter. A successful international firm must be able to operate over long distances, having the flexibility to allow a high degree of independence to field operations very far away from the home office. If your firm is stretched thin by the press of business at home, and you are not prepared to devote a major block of management time and talent to an international effort, stay home.

Above all there must be the understanding that things will be different. Markets in the "developing countries" (we'll still use that word for the want of a better descriptive term) are different... just how different becomes evident when the hunt to explore business opportunities abroad begins.

Those who have ventured into the global market and have found success, will all agree with a few general statements: It will take longer than one thinks. It will require much more administrative time and effort than anyone thought reasonable. And the final result will be far different from what was visualized.

THE INTERNATIONAL MARKETPLACE

Not all international markets offer equal opportunities. Selecting and evaluating a "target" country is the first, and probably the most important, task facing any firm that finds itself ready for international activity.* For those companies that have a product to sell, the assessment of market opportunities is rather straightforward; however, for those in the energy service business, where service over time is the key ingredient, market assessment is a "whole different ball game."

---

*In chapter 19, "The Global Perspective," Bob Dixon offers a valuable glimpse of the international marketplace.

Some years ago, our firm developed a protocol to help evaluate the opportunities for an ESCO to do business in a given country. Over time, the protocol has been tested, refined and proved itself useful in assembling the data needed to make reasonably sound decisions. After working in more than 30 countries, we are convinced that no list can garner all the needed information or remove all the risks. But our protocol removes many of the risks resulting from wrong assumptions. The format, which we have evolved, can be roughly divided into three broad areas: country analysis, in-country partner potential, and market assessment.

COUNTRY ANALYSIS

For a firm which is serious about entering the ESCO market in a certain country, there are several categories of information that deserve attention; some quite obvious and some rather subtle. While overlaps may occur, these general areas of inquiry provide a way to cross check data and help with interpretation.

Some of the needed information is readily, and publicly, available … and may even be accurate. Many other key pieces of information need to be developed in a visit to the country in question or through someone designated to conduct research on behalf of the firm. Reputable consultants are available who have the contacts and the knowledge of international business to get the answers and develop a plan forming the basis for solid decisions. Needless to say, it is beneficial if the consultant has an understanding of the energy service business, so that the information generated is truly germane.

ECONOMIC ISSUES

First, of course, a careful examination of the economy is crucial to long term contracting; so it is essential to assess the overall economy and the political situation of the country in question and understand how the economy meshes *with* the political situation. An initial reading on the stability and philosophy of the government and, in the case of a number of countries, the level of privatization, can be relatively easy to obtain. Much of this can be done "long distance" from sources that are quite readily available. In the United States, the U.S. Department of Commerce provides a constant flow of reports on the economy of most nations. This information is very good, although, with some exceptions, it tends to be quite general and slightly dated.

The U.S. Chamber of Commerce maintains a group of "Business Councils" consisting of business people with involvement in various countries. They publish valuable reports and provide the opportunity to meet with officials from their subject countries and other American businesses working in these countries.

The World Bank and the other multilateral financing institutions also publish a great deal of material, which often offers valuable economic analyses. These sources are all useful; however, time and effort are required to sort out the useful from the interesting (or boring).

Beyond this broad economic assessment, there are a number of more subtle "blanks" that need to be filled in: What is the *real* posture of the country's leadership toward trade, the economy, foreign-based business activities, energy, etc.? How stable is the political situation and are there important changes in the offing? What are the prospects over the next several years? Specific economic issues, such as inflation and interest rates, convertibility of currency, repatriation, and economic trends are all indicators of possibilities/limitations for business success.

The answers to these questions can only tell the prospective international ESCO whether the target country MIGHT be a possible market and MAY be worth further effort and expense. To use the parlance of the ESCO industry, this serves as a "scoping audit" to see if the expenditure of more time and money is warranted and how the country in question compares with other opportunities. Securing further data, beyond the broad facts suggested above, becomes more costly, and the data more subject to interpretation. Each of the initial "facts" opens the door to more issues, which will require careful examination and expansion, before intelligent business decisions can be made.

There are other things about the government, the energy situation and the economy, that data from the U.S. Commerce Department and other sources will not tell you. Does the government policy encourage energy efficiency or do various rules and regulations make it very difficult to get anything done… if you are looking at the market in public facilities, this is a vital question. Public espousal does not always match the facts and "policy" may be very far from reality.

This essentially seamless inquiry leads to more basic questions: If you go into a country, what kind of a market will you face? What problems will you have to deal with? What resources can be called upon to get the job done? At this stage there is only one certainty; *it will not be like home.*

LEGAL ISSUES

The legal systems in possible target countries may not yet be ready to handle some of the "everyday" matters that the U.S. ESCO industry takes for granted. Laws may be solid, but the time frame in which the legal process works may make contract enforcement nearly meaningless. The legal framework in India is solidly grounded in English common law, but the courts are so slow that almost anything can happen. In some transitional economies; e.g., Eastern Europe, Russia, China and others, the court system, as we would recognize it, barely exists. Contracts can be more than somewhat questionable.

Efforts are being made to build a solid court system in most of the transitional economies and it will get better. It will take time and progress will be very uneven.

ESCOs which were in business in the early 1980s, are apt to remember the problems with U.S. attorneys rather fondly once they try to explain a performance contract to attorneys in some countries. An important first step is to present a contract to an in-country attorney; then, discuss the reasons for the various provisions with the attorney. Allow the attorney to fit it to the local laws; then, critique it to be sure the ESCO is protected and the potential customer's position is reasonable, keeping in mind various aspects of contract enforcement. For example, specific information as to product value can greatly facilitate court action in some countries. In any case, the contract you will ultimately use will have little in common with the (more-or-less) standard contract that most ESCOs use at home. With careful work, the key provisions will be there but the form may be hard to recognize.

BANKING ISSUES

Banking in some of the countries that seem to be attractive targets can be a problem. Commercial banks in some transitional economies may still be in an organizational phase and not really ready to consider the financing of long term projects. In other areas, a "long term" loan is two years or less... and this may be justified in an economy where the inflation rate is high. Through the support of the World Bank and USAID as well as independent efforts, some work has been done toward educating the banking community in a number of countries in the theory and practice of performance contracting, including the shift in thinking from asset based financing to cash flow financing. The impact and effectiveness of this effort varies widely, but the "seeds" are being sown.*

In countries where a fledgling ESCO industry has begun to develop, some banks have taken note of the opportunity and funding has, gradually, become more available. As in all matters, the variation between countries is great and no generality will answer the important question: Is financing available in-country? And is there the possibility of accessing external funds? There are other sources that will fund US ESCO work outside the US, especially where it involves export of American equipment; e.g., US Export-Import Bank. All of this suggests that it is important to know where you can get financing for your efforts or where your potential customers can get credit before you make major commitments.

In nations where the multi-lateral development banks (MDBs) have been interested in energy efficiency, other routes to project financing may be available. MDBs can support/educate and/or mitigate some risks through their involvement. The European Bank for Reconstruction and Development (EBRD) has taken the MDB lead in energy efficiency financing and moved with considerable success in parts of Central and Eastern Europe. The World Bank and the Asian Development Bank have encouraged significant progress in China and some other Asian countries. The World Bank's effort to create and maintain energy efficiency financing dialogue between China, India and Brazil offers a great model for international discourse. The Inter-American Development Bank is active in Central and South America and USAID has worked to encourage energy efficiency in countries around the world. There may be ways in which these resources can be tapped for specific projects.

HOW "FOREIGN" ARE YOU?

Underlying suspicion or bias against the entrance of foreign businesses, investment or technology can be very real regardless of official

---

*Please refer to Chapter 18, "International Financing," by Thomas Dreessen for a discussion of the work of the International Energy Efficiency Financing Protocol.

policy and what is published. It pays to look beneath the surface. If suspicion of foreign investment exists, the bureaucratic road blocks, all quite unofficial, can be almost limitless. In India, for example, cultural or regional differences and/or political practices have left the states quite independent of the central government. The Government of India can welcome a foreign business with open arms, but that may provide no help at all in a particular state. A classic illustration was the attempt by Enron and Bechtel, with the enthusiastic support of the Indian Government, to build a power plant in Maharashtra state on the Indian coast. The state was in the midst of a political fight with the Central Government. And, although many millions were spent on construction, the plant went almost nowhere for years. Workers were harassed and supervisors were arrested from time to time. The Enron group gave up and swallowed the losses but the plant eventually did open ... and closed in 2001. Seems the local State Electricity Board felt the rates charged were too high. There is, at this writing, a plan to bring the plant, India's largest ever foreign investment, back to life. Maybe. All this in a country desperately short of electric power.

It is not enough to know the "national policy" on the acceptance of foreign companies. A look at what happens to foreign companies as they work to get established can tell you a great deal more.

ENERGY MATTERS

Knowledge of the overall energy "climate" in a target country is key to good planning. Moving beyond "internet available" information, there is very important energy information to be garnered "on the ground," such as power availability by region and power quality. In thinking about energy efficiency projects, power shortages in a particular region and frequent outages are critical concerns. A problem or an opportunity? In India, power outages and wide power quality swings are facts of life. In the Philippines and some Eastern European countries power curtailment is a daily, scheduled, occurrence. Negotiations will undoubtedly be required to determine how these "routine" events affect savings calculations. Cogeneration, distributed generation and/or standby power may be important ESCO offerings.

A growing number of countries have horizontal stratification of utility services like the U.K. model; i.e., generation, transmission and distribution, as separate functions, with portions of the business controlled by the state. A key question arises: Who loses when energy efficiency increases? A utility that believes its revenue will fall if a customer's energy efficiency increases can find a lot of ways to make things difficult even if that utility's official policy encourages efficiency. It is important to understand utility tariffs. If increased efficiency leads to a substantial rate increase the viability of a project comes into question.

A careful look at energy pricing, the impact of government subsidies by market segment, and general pricing trends are basic to the potential for economically viable projects. Utility structures, policies, and attitudes vary by country, sometimes by region and almost always by utility company. The incentives and/or disincentives that originate with a utility can make a huge difference in the possibilities for successful energy efficiency projects.

How is energy purchased? In Poland where district heating is a big factor, tenants in apartment blocks historically paid for heat by the square meter; thus, they had no incentive for energy efficiency. Without metering, there was no incentive to save. If a "no payment crisis" exists, such as the situation is Russia, there is no incentive to use energy more efficiently. As a professor at The Moscow Power Engineering Institute put it, "When you are not paying, it makes no difference how much you are not paying." These and other factors have complicated the approach to energy efficiency marketing and have required some very innovative strategies.

In some countries there is very little accurate metering. Even simple consumption meters may be scarce. There may be demand schedules, but no time-of-use metering. One very large Indian utility knows when its peak loads occur, but does not know exactly which customers are pushing the top of the scale. Their solution to potential system overload is to call some of their "high tension" customers and ask them to cut their loads.

Due to the chronic, and growing, power shortfall in some countries, manufacturing companies have contracted for far more electricity than they need as "insurance" against the time when they may be asked to cut back consumption, or when they want to expand operations. This can well mean a minimum monthly bill that cannot be changed when energy is saved, without a difficult contract re-negotiation. This poses a potentially serious disincentive for an energy efficiency project.

The problem of stolen power plays a role somewhat beyond our usual thinking. While presenting a seminar on ESCO business operations, I was talking about collecting savings-based payments from customers when the question was asked, "How do you collect from a customer who uses mostly stolen power?" Finally I managed an answer, "Well, if the customer is stealing his power, he probably won't be very interested in energy efficiency in the first place." Not brilliant but perhaps accurate. Stolen power is not at all unique and can be a problem in several parts of the world. It can also corrupt the data used to calculate savings and resulting payments. Energy intensity data across an industry can be meaningless.

Gathering as many indicators as possible as to what is likely to happen in a country is key. Some guesses may not be entirely correct, but they can be important; and, in conjunction with other data, are revealing. Are new taxes in the works, or are there tax breaks that could make a difference? What sectors of the economy are growing or shrinking and what sectors are being propped up by government at an artificial level? Will those props last? And, not incidentally, how do people feel about the economy? If people are optimistic, they are far more apt to be ready to accept new ideas and approaches. The optimists are much more apt to take advantage of energy opportunities. Perception often becomes reality.

In-Country Support

If an ESCO is going to do business in another country, it needs to know what resources are available in terms of technical capabilities, business alliances, financial resources and marketing support.

In many transitional and developing countries, good technical people are in abundant supply. Technically capable companies, which are willing to work, and are well qualified to be involved in energy efficiency projects, are plentiful. However, the "software," the knowledge about the business side of energy efficiency, and more particularly, knowledge about energy performance contracting, is usually lacking. The concept of full service energy service companies is growing, but to many it is still new.

In some countries the level of technical capability is more limited. In many instances, the "new" equipment they are most familiar with is of the '50s era. In China, chain grate boilers are very common. If the reader is not familiar with that type of equipment, it is not surprising-they disappeared from most of the world at least 50 years ago. (The Chinese have developed a couple of ways to get their efficiency up around 50% but...) An ESCO must be prepared to deal with old technologies that the owner intends to use for some time into the future.

Misinformation among the "informed" can also be a problem. Disassociating a utility DSM program from performance contracting can at times be a struggle. While the concept of a full service ESCO is readily accepted, the business approaches we learned almost at birth are new and often thoroughly misunderstood in some cultures. The language barrier can be far greater than the difference between English and Hungarian. Terms such as "return on investment," or even "profit" are still surprisingly new in some of the former communist countries. The words may be part of the lexicon, but it is a mistake to assume that such terms, or their long term implications, are well understood. Some times they talk the jargon, use the business terms we are familiar with, but their ability to actually implement the concepts is seriously narrow and lacks a full understanding of economic impact. Understanding the technical and business sophistication available within a target country is one more criteria to weigh before major decisions are made.

Very good, very astute business people can be found in any country; and, in those countries where the economy has been unleashed and given a chance to respond to markets, their numbers are increasing. The rapidly growing numbers of young people who have been exposed to the market economy and have grown up in a less regulated society are making a significant difference.

A WORD ABOUT CULTURE

Much has been written about the importance of understanding the cultural climate in a country in which you wish to do business. It is very important. It is equally important to realize it cannot all be learned, *absorbed* and understood instantly. It takes time. Time, which you will be reluctant to spend especially when demands, such as evaluating potential markets, call to you.

The initial solution is a willingness to learn and an acceptance that Beijing, Tula, Warsaw, and Ahmedabad are not like home. This understanding, coupled with a degree of reserve and a willingness to be flexible, can carry you a long way. An official in Moscow commented, "It is easier to deal with Germans than Americans, they are more flexible. Americans think theirs is the only way to do things." For those of us who have always viewed the German approach as rigid, this comes as something of a shock, but some thought suggests a reason. Germany is surrounded by differing cultures and to do any business outside of their own country they have had to recognize differences and be sufficiently flexible to get along with their potential customers. Most Americans are rather new at this game.

As an aside, it is well to remember it is THEIR culture. You are not there to change it. You must work in their accepted way of doing business. Any attempt to do otherwise is an invitation for disaster.

In some countries there is a regular ritual for the presentation of business cards. And in some cultures business discussions have a defined time and place. Some gestures that would pass without notice or comment at home can create real problems. As an example, a dedicated alumni of the University of Texas runs a real risk if he flashes the "Hook 'em Horns" sign in certain countries such as Italy. It can be a "fighting" insult. Our finger and thumb sign for OK could get you flattened in Brazil.

Words may have meanings we have never thought about. In India, describing a woman as "homely" is not insulting, it just suggests that she is good at household management. (But do not try this on an Indian hostess. She may have been educated in the United States). Even your wonderful company slogan may suffer in translation. A Coors beer line used for a time, "Let It Loose," in Spanish read, "suffer from diarrhea," and probably did not sell much beer. A number of large companies, who are deeply into international trade, routinely run their new product names through a translation screening process just to be sure they can avoid embarrassment.

Don't try to tell jokes in English that have to be translated. With very few exceptions, they will not translate into anything remotely funny... and puns simply won't work at all. If your interpreter speaks pretty good English he, or she, may laugh a little, but may, as happened in one case, simply say to the "audience," in his language, "Laugh now, Mr. ____ has told a joke."

Negotiations in another culture can be a disconcerting experience. In most cases negotiation of a contract in the U.S. begins with the assumption that both sides should emerge with a result that is fair and equitable. There is an assumed level of trust. In many other parts of the world, negotiation begins with an assumption that neither side is very trustworthy and the process tends to be far more competitive with touches of drama. This has probably grown from the age old tradition of bargaining in the market place and should not be taken as a reflection on your integrity but rather as just their way of approaching business.

The ways in which cultures differ cannot be learned instantly. Advice and guidance of someone totally familiar with local custom can, and will, be extremely valuable. So approach a new culture with interest and curiosity. Don't rush things, watch your hosts and learn. They will understand and appreciate your efforts, and will be inclined to overlook your mistakes. Verbal recognition, on your part, that procedures, which worked well some other place, will need to be modified to fit local conditions will be well received.

A final "culture" caution: Even if you feel pretty good about your knowledge of the local language, don't try to negotiate or do serious business in a language you are still learning. There are just too many nuances that you will not know yet and real misunderstandings can result. Use an interpreter to be sure you know what you are agreeing to.

FIND THE TIGERS

In many countries, it will be possible to find a few "tigers"… people who immediately grasp the concept of full service ESCOs and performance contracting; people who have enthusiasm and drive and who see a personal, and/or national, advantage in making the concept real. Some of these people will have the power and the will to make things happen. These "tigers" may be in government, highly respected industry or financial leaders, or an eager young engineer just getting a good start. The existence of a tiger can help to balance other concerns, especially when it comes to getting around bureaucratic roadblocks or opening doors.

In China, Mr. Li, then Director of the Energy Division of the State Economic and Trade Commission, attended a seminar on performance contracting. He asked questions (all translated from Chinese) and made comments. During a tea break he was enthusiastically explaining financial concepts to other class members, drawing furiously on a blackboard. We had found a tiger. Together with a young woman from a commercial bank, who spoke English well and was every inch a tiger in her own right, they became invaluable in forging the creation of Chinese ESCOs.

IN-COUNTRY PARTNERS

Few ESCOs will make it in a foreign market without in-country help. A joint venture with an in-country partner, or at least a strategic alliance with a well established firm, or a well established branch of a company in a related field, is a must. The one exception may be large multi-national company that already has a presence in the targeted country. A careful census of who is in the energy business and what they are doing is an important part of the data collection process.

It is fairly simple to find out who is in the energy efficiency business within a given country. Engineering or mechanical contracting firms, who do turnkey energy projects and in some cases, savings-based projects, exist almost everywhere. In addition, thanks to the efforts of the U.S. Agency for International Development, quite a number of consultants offering energy audits are available. (A word of caution: they may not be doing anything like an investment grade energy audit.) These may be likely competitors, or present opportunities for joint ventures or alliances. At the very least, they are apt to be the people an ESCO will need to count on for installation and operational help.

Also be aware these firms may see a partnership with you as a way to learn your business and once done, cut you out. The trick is to be sure your expertise and capabilities continue to be vital to their own future success.

Discussions with a few of the firms in energy related businesses will quickly reveal what is happening in the efficiency market. But, be aware: *their definition of what an ESCO is, and what it does, is apt to vary tremendously from US view.*

There is one more possibility for an in-country partnership that is well worth examining. There may be a bank or financial house that can see the advantage of helping its own customers cut their costs through energy efficiency. They have the advantage of knowing a great deal about their customers and, if they understand the concept of performance contracting, will can see how the bank can benefit. We have found some bankers who immediately grasp the potential, others …

The criteria our firm has established as a way to judge the qualifications of a potential in-country partner are not very different from what one would use to select a partner close to home … with some important variations. Along with the technical capabilities and the strength required for project implementation and follow through, there is a critical need for a partner that has the contacts and the knowledge to navigate the regulatory maze of that country. In-country partners also need to know how to deal with the codes and standards that are the pride and joy of any bureaucracy. Finding the way through the government channels is not usually considered an engineering talent; so administrative ability is also an important consideration. This talent will frequently go beyond *what* the in-country partners know, to WHO they know.

An effective in-country partner will also know where, and how, to acquire and/or import material and equipment needed to move projects forward. Tariffs may be high, but there may be "forgiveness" of tariffs for some types of energy equipment... if the partner knows how to work the system and use the rules to their advantage. In one country, we were told several times that it was almost impossible to take advantage of the low tariff rate for importing energy efficiency equipment, then a young engineer said, "It's easy if you know how to write the description of the equipment you want to bring in."

Local knowledge of the availability, and quality, of sub-contractors who will get the job done is also a vital resource, which cannot be brought from home. In the final analysis, the in-country partner needs the know-how and the contacts to get the job done efficiently and effectively.

Market Assessment

Selecting a target market within a new country, and identifying opportunities within that general market, is far more complex than in the US. The abundance of statistics we take for granted in the United States, and in most Western industrial countries, is often rudimentary or non-existent in many countries. This lack of data extends from a good estimate of the energy requirements relative to production levels, to the credit information that financial people want to talk about. The information generally exists *somewhere*, in some form, but it will often require extensive research in-country to get it. It is seldom available in neatly written form and it is very frequently suspect. Historically, countries, which existed on imposed "five year plans," frequently reported statistics to satisfy plans and not to reflect actualities. A tendency still exists to make all forecasts look rosy and all results outstanding. Healthy skepticism makes good business sense.

In countries where business was kept under tight reign for decades, there is an understandable reluctance to "report" what they consider proprietary data. Questions such as "Why do you need to know...?" or, "What are you going to do with this information?" are not uncommon. Although perhaps different in degree, these questions are not very different from what is asked by potential industrial customers at home. Diplomacy and perhaps some reassurance from an in-country partner can usually bring out the information you need although occasionally an in-country partner may have more difficulty getting answers than would a perfect stranger.

An ESCO, contemplating markets on an international basis, should first consider where its own strength lies. If the firm's management knows a lot about a specific industry, such as the cement industry, hospitals or office buildings, that is the place to start. The firm is better equipped to assess the data, the problems, the opportunities and talk the language.

A few key indicators are basic in determining whether a particular industry or market is worth consideration. Preliminary screening includes concerns as to what industry segments are apt to be viable over time. The state of privatization in that industry and/or the level of government subsidies for the industry are key issues. Many of the industries in transitional economies, which have not yet been privatized, are "buggy whip" manufacturers that will never survive on their own. Not good targets for long term energy efficiency projects.

The "flip side" of the privatization issue involves those facilities, which are apt to remain under state control. They may be quite viable as targets for energy efficiency. These are public institutions, such as hospitals, schools, prisons and other public facilities as well as industries "vital" to the government. In fact, these markets initially may present the most stable, least risky opportunities. The political and economic decisions have been made; they will be there over time.

Is an industry growing or is it declining (or perhaps dying out)? Who is investing in the industry, or in specific companies within the industry? If outside money is flowing in, it suggests that someone else has confidence in the long term viability of the industry. Within an industry, companies that have attracted substantial outside equity participation may be particularly attractive targets for energy efficiency work. Their management is apt to have a better appreciation of costs as well as the importance and advantages of increased efficiency.

A real "deal killer" in market selection can be the question of whether potential customers in that market can keep their savings. If savings flow to another entity, neither the financial incentive nor the means to pay the ESCO exist. Before a potentially viable market is written off, however, it pays to look at the entire revenue stream.

As our firm brought together the parties to establish the first performance contract in Eastern Europe, all conditions seemed very positive until we discovered that the Bulovka Hospital in the Czech Republic could not keep its savings. Another means of accessing the revenue had to be found. After considerable research and discussion we were able to access the National Insurance Fund which finances hospital operations. A way was found for the hospital to benefit from energy savings. The project went forward, and all parties, Bulovka, Landis & Gyr, Energy Performance Services, and Kiona International were all proud to take credit.

TEST THE "FACTS"

All "facts" accumulated to examine market opportunities should be tested... a source that is *absolutely sure* of his/her knowledge of a market, very probably is not. Asking more questions and/or seeking other sources is critical. Generalities, which are readily available, can hide real problems or opportunities. It is easy to say that country "X," with a very high energy intensity, is a good prospect for energy efficiency; however, that bit of information does not tell you that the government bureaucracy makes it almost impossible to get anything done. Or, that energy users don't pay their bills. Much as a child stacks building blocks, each accumulated fact should contribute to the picture of a potential market. The stronger the building block base, the easier the decision to take it one step further. But test at every level.

There is a wide world of opportunity for an ESCO that takes the time to do careful, thorough "homework." It is necessary to accept from the start that doing business in India, China, Slovakia, Brazil or Ukraine will not be the same as doing business in Kansas or California. Only one assumption works: assume that most assumptions about those countries are wrong. Good people can be found in every country, who are eager and well qualified to act as effective partners. Finding them is the challenge.

An in-depth investigation of selected countries with limited risks can allow an ESCO to bring together the ingredients that can lead to success. An ESCO, which does not have the internal resources to make a careful business evaluation in a selected country, should secure outside consultation or stay home.

Chapter 18

International Financing

Over the past ten years, hundreds of millions of dollars have been invested in international financing energy efficiency projects (EEPs) and the development of energy service companies (ESCOs). Despite the work of numerous international agencies including USAID, the UN Foundation or the Global Environment Fund, and multi-lateral development banks (MDBs) including the World Bank, the International Bank of Reconstruction and Development, and a host of others, very little progress has been made in removing project financing as the single largest barrier to the widespread implementation of energy efficiency and the energy service company (ESCO) industry around the world. This in no way is intended to be a criticism of the intent of any of the international agencies or MDBs, but rather to point out the difficulty in creating viable financing structures for energy efficiency.

The investment made by several MDBs has focused on developing guarantee mechanisms for local banks and financial institutions within targeted developing countries. Probably the most successful program to date is the International Finance Corporation's guarantee mechanism in Hungary, which has resulted in one of the more robust performance contracting industries in central Europe.

One of the most unique investments was made in China by the World Bank, Global Environmental Fund and International Bank of Reconstruction and Development to establish and fund projects for three Energy Management Companies (same as ESCOs), totaling over $150 million. The primary goals of this investment were to introduce, demonstrate and disseminate the advanced market-oriented energy conservation mechanism of "Contract Energy Management" (same as "Performance Contracting") and to strengthen effective dissemination of energy conservation information in China. During the period 1997 through 2004, the three Chinese ESCOs implemented and funded 346 projects for a total investment of over $100 million.

The goals are generally believed to have been accomplished, and from a performance contracting industry perspective, the program should be highly applauded for its promotion of the savings-based business model.

Unfortunately, this program only provided government-backed project funding exclusively to the three ESCOs, and therefore did not make any type of financing available to the rest of the Chinese energy efficiency industry and did not create any commercial financing mechanisms within the Chinese local banks and other financial institutions. Consequently, the rest of the Chinese ESCO industry is now struggling to grow (and sustain) because of their inability to access meaningful financing for their projects on a "paid-from-savings" basis.

The World Bank has recognized this barrier and has attempted to overcome it with a $25 million loan guarantee fund. Established in 2002, the fund was expected to leverage $250 million in lending from commercial banks for ESCO projects. However, this guarantee program is generally regarded by the Chinese ESCO industry as a failure because it requires them to provide personal or other assets equal to the loan value to the guarantor as security for their energy efficiency loans. This allegedly has prompted some owners and managers of ESCOs to provide their personal residences as collateral.

International agencies have made substantial investments in developing countries intended to create and develop ESCO industries. A lot of the funding was used to pay independent consultants to create and conduct technical training workshops for local engineering, construction and other types of small consulting and contracting companies within the target countries to become ESCOs. The training programs focused primarily on teaching them how to develop and implement savings-based energy efficiency projects (EEPs) and included such things as how to perform energy analyses, quantify savings opportunities, create energy baselines and related savings measurement & verification (M&V) plans. These programs also tried to teach the commercial and financial skills needed to apply the performance contracting model and become an ESCO.

A Need for Commercially Viable Project Financing

While this capacity building for ESCOs in nascent markets is truly appreciated and a testimony to the global appeal of the performance contracting business model, it also has created negative perceptions about ES-

COs resulting from their limited success in implementing any meaningful level of EEPs. It is particularly disturbing to hear this criticism about these failed ESCO efforts because in almost all cases, it was caused by little or no development funding being applied to create commercially viable project financing for the ESCOs. Without COMMERCIALLY VIABLE project financing, the ESCO industry will fail because it cannot deliver its "paid-from-savings" business model. Trying to develop an ESCO industry without project financing is analogous to trying to develop a banking industry without providing money for the banks to lend.

When I first started developing performance contracting projects on an international basis in 1994, I thought it was the responsibility of the MDBs to solve the financing barrier for EEPs that existed then and, too often, exists today. However, I no longer think this, but rather feel the role of the MDBs is to "fill the gaps," enabling local financial institutions to provide the financing. MDBs are not the proper entities to provide the on-going, sustainable solution to this international financing problem, and it is not their responsibility to be the primary provider of financing. They should continue to provide development funding support to local financial institutions (LFIs) and help create new EEP financing mechanisms for local banks and financial houses to implement in the marketplace.

So where does that leave us regarding the status of international financing for EEPs? Generally speaking, there is a complete lack of readily-available or commercially attractive financing for EEPs in most international markets. End users, manufacturers, contractors, developers and especially ESCOs encounter significant difficulties in financing even the most cost-effective EEPs. This lack of viable financing is not only a tremendous deterrent to the widespread implementation of worthy projects, but also to the creation of robust energy efficiency industries. This is especially the case with the ESCO industry due to its total reliance on the availability of financing to implement the performance contracting business model.

Any viable solution to the project financing problem must include the availability of financing from local sources with the loans to be repaid in local currency. As previously stated, it is not the mission of the MDBs to be direct lenders in support of EEPs and even if they were so inclined, it would not work. These individual projects are small and typically installed in privately-owned facilities… conditions that normally would not meet the due diligence requirements of an MDB, nor meet their typical requirement to repay in hard currency. It is clear that local financial sources within each country or economy need to be the providers of financing for individual EEPs.

Contrary to the belief of many, in most international markets there is plenty of funding capacity in the local financial institutions. The problem is not due to a lack of available funds, but rather, due to an inability of end users, manufacturers, contractors, developers and ESCOs to access such funding capacity for their EEPs in market-acceptable ways. This inability is caused by a fundamental "disconnect" between the traditional asset-based lending practices and the savings-based project financing structure needed to motivate end users to implement energy efficiency projects to be paid from savings. This "disconnect" is exacerbated by the relatively small size and complexity of EEPs, which discourage local financial sources from venturing into "new territory." Too many banks and other financial house have yet to invest in the development of an internal capacity to understand the risks/benefits associated with the financing of EEPs and have not created new financing structures needed for the energy efficiency market place.

Several of the common challenges encountered by ESCOs and other developers in trying to finance EEPs locally in international markets include:

- *Local financial institutions are primarily accustomed to providing "asset-based" lending,* which typically limits the amount of debt financing provided to 70%-80% of the market value of the assets provided as collateral by the borrowers. Experience has shown that the cash flow values generated by efficiency savings are not considered in the credit analysis.

- *Local financial institutions are not familiar with the intricacies of financing EEPs* and do not have the internal capacity to properly evaluate the risks and benefits of their design, nor to structure financing in market-acceptable ways.

- *Local financial institutions do not properly value the positive cash flow generated by EEPs as a new asset for end users.* They do not fully understand (or believe) that these projects will generate new and sustainable positive cash flow from reductions (savings) in the end user's existing operating expenses. Also they are not comfortable with the idea that, in many instances, the savings are sufficient to repay project loans, thereby not requiring the end user to have to repay the financier from existing operations. It is very difficult for local financial institutions to view savings gained from efficiency improvements as hard assets with enough market value for them to feel adequately secured.

- *Lack of familiarity with the potential benefits of increased energy efficiency within the local financial community creates a perceived (and unwarranted) high-risk lending profile for projects,* resulting in commercially unattractive financing terms and conditions.

- *Local financial institutions are unwilling to invest the time and resources needed to develop a new lending product, such as the financing of EEPs.* As previously stated, this is due in large part to the relatively small size and complexity of these projects, and the uncertain market size of the energy efficiency industry.

- *Local market conditions are frequently not conducive to even allow LFIs to offer commercially-viable financing terms and conditions.* Examples of such market conditions in developing countries include high interest rates and short repayment terms of 3 years or less. These conditions make it very difficult, if not impossible, to structure project financing so as to be repaid from savings. Unfortunately, these conditions are often caused by unstable currencies, inflation, regulatory and other macroeconomic events outside the control of the local financial institutions.

- *Local financial institutions frequently fail to see the financing of EEPS as an added business opportunity,* and a new service they could offer their clients.

The above challenges to financing EEPs can be significantly reduced on a global basis through the implementation of a two-pronged approach that comprises: a) CAPACITY BUILDING within the local financial structure, and b) the creation of MARKET-BASED INCENTIVES.

CAPACITY BUILDING WITHIN LOCAL FINANCIAL INSTITUTIONS

While the commercial and legal aspects of financing EEPs are largely specific to the local conditions of any country or economy, the special intricacies that are fundamental to financing such projects are not specific to each one. There exists a critical need within local financing sources to develop a better understanding of how a project, which increases energy efficiency, can generate new cash flow from existing operating expenses; thereby increasing the credit capacity of end users, which in turn, enhances their ability to repay the related loan.

To meet this need, training should be provided to local financial institutions on standardized techniques and procedures (a "protocol") for financing energy efficiency, which would show them how to evaluate the risks/benefits and create the needed structures.

AN INTERNATIONAL ENERGY EFFICIENCY FINANCING MODEL

The development of such a protocol would be a "grass roots" effort, similar to, or in collaboration with, one already under way by the Efficiency Valuation Organization (EVO), formerly IPMVP, Inc., in its development of the *International Energy Efficiency Project Financing Protocol* ("IEEFP"). The goal of the IEEFP is to become a global "blue print" for educating and training of local financial institutions in the financing of energy efficiency. At the same time, IEEFP could create standardized M&V and financing protocols that will enable aggregated financing and reduced transaction costs. Such a protocol could also create a platform for standardized M&V protocols to be used in measuring CO_2 reductions for CDM projects. The approach envisioned by EVO contains a consensus building process, which will be iterative in nature and involve relevant stakeholders within a country or economic region, as well as accessing global financial, business and energy experts with IEEFP experience.

Some of the major components contained in the IEEFP are listed below.

- Criteria for projects to qualify for financing include but are not limited to:
  - Minimum project economics, such as payback criteria, coverage ratios, etc.
  - Pro forma cash flow formats
  - Acceptable technologies
  - Acceptable technical and financial risk profiles

  - Construction payment terms
  - Acceptable savings measurement and verification (M&V) protocols
  - Capabilities of companies implementing and monitoring EEPs;

- Procedures to evaluate financial and technical aspects of EEPs that include:
  - Estimated savings and construction/implementation costs
  - Savings calculation and baseline methodology
  - Savings measurement and verification protocol
  - Environmental aspects
  - Technology risks
  - Construction progress review
  - Completion and commissioning requirements

- "Lending Criteria" for EEPs by borrowers;

- Credit information required of borrowers;

- Credit review and analysis procedures to be used by local bank management for loan approval;

- Standard terms to be included in loan documents, security agreements, contractor agreements for installation, maintenance and M&V;

- Written training manuals and case studies for use to train personnel at local financial institutions on how to qualify customers, negotiate and execute all loan-related documents as well as properly securing the collateral and other pertinent matters;

- A risk sharing approach which employs formal risk assessment and mitigation techniques, develops one or more payment security mechanisms to secure the cash flows from energy savings for debt repayment, and results in greater confidence to LFIs regarding their assessment of customer and ESCO credit worthiness.

- Standardized evaluation and approval parameters for small projects in general that mitigate risks on a portfolio basis.

Market-Based Incentives

A protocol, such as IEEFP, alone will most likely not be enough to achieve implementation of EEPs on a meaningful scale. These projects are generally viewed by private end-users as a low-priority "infrastructure" investments versus an investment in the growth and development of their core business. Unless energy efficiency projects can meet very aggressive hurdle rates providing simple paybacks of less than 2 years, they cannot compete for the limited capital of private businesses.

These barriers, plus other complexities associated with EEPs, such as the measurement of savings, create a somewhat apathetic attitude by end users and financial sources. Financial incentives are, therefore, needed to accelerate the adaptation of an IEEFP-type model, which can in turn help to achieve the more important goal of widespread implementation of sustainable energy efficiency.

In order to address the unique regulatory, legal and commercial conditions of individual countries, financial incentives are apt to be different for each country. Any successful financial incentive program, must provide additional economic value and mitigate risk (perceived and real) to properly motivate both end users and the financial community. The incentive options are many and could include a variety of things such as:

- Below-market interest rates, or other similar financial benefits for end users to motivate them to implement and finance energy efficiency;
- Guarantee mechanisms to local financial institutions to cover a portion of their potential loan losses, thereby reducing their risks;
- Financial Instruments that cover a local financial institution's exposure in providing extended repayment terms required to allow for reasonably sized EEPs to repay financing from operating savings;
- Additional profits paid to banks on energy efficiency project loans so they can make more profits on financing them versus their traditional loans;
- Making potential customers aware of the EEP financing opportunities;

- Funding of resources to train the staff of local financial institutions about the IEEFP and to create the required lending infrastructure;

- Any and all other financial incentives available in the markets that permit local financial institutions to offer commercially viable and attractive financing terms for EEPs.

As noted above, incentive options might include a variety of things like buying down the cost of capital on the EEP, extending financing terms, guaranteeing portions of loans, etc. One mechanism for funding these incentives could be the establishment of an EFFICIENCY BENEFITS FUND, which is analogous to the "Public Benefits Charge" concept in the U. S., Brazil, Thailand and other countries. This mechanism represents a sustainable funding source from the ratepayers and has been quite successful at accelerating the implementation of energy efficiency in several states in the U.S. including New York, Massachusetts and California. However, it would clearly require regulatory and policy support in countries, which do not currently have such mechanisms.

It is important to note that no matter what incentives are developed, they are only to be provided to *supplement and not to actually finance or fund* EEPs. The actual financing needs to be underwritten by the local financial community in order have an infrastructure created, which will sustain over a long time.

Chapter 19
Globalizing an ESCO

WHY GLOBALIZE?

Why would any ESCO in their right mind embark upon the challenging journey of globalizing their business in lands far far away? There are two primary reasons, one being growth, another being responsive to their multi-national customers, who are asking them for help in addressing the many needs and issues that customers face in today's business environment. We all know that businesses have to grow to survive, and have to respond to customer needs in order to grow. Whether it is the desire for growth, or responding to a customer's needs, which is driving an ESCO to globalize, the challenges remain the same. How do you "export" the ESCO *"product and services"* to a geographical market that will not have the business, legal, or financial infrastructure in place to support the effort in a yet-to-be-created market?

There probably exist many views on what an ESCO is, but it is important to understand my definition of an ESCO to help understand why globalization of their *"products and services"* will be challenging.

WHAT IS AN ESCO?

ESCO's have eight key competencies: customer acquisition, project development, energy engineering, project management, financial engineering, risk management, service, and performance assurance. ESCO's, using these eight key competencies, custom design and through turn-key implementation, deliver major building technology, energy efficiency, infrastructure, and operational improvements with guaranteed financial and performance results, while having a positive impact on the environment.

Noticeably absent are defined technologies, the requirement to manufacture technology, vertical market focus, or the requirement to create and sell energy. Some might say that this definition is so broad that anybody can be an ESCO. My view is different. I think to be an ESCO you must do it all, not just one narrow slice of the definition.

WHAT ARE AN ESCO'S "PRODUCTS AND SERVICES"?

ESCO's have developed *"products and services"* to meet a wide range of customer needs. In my global travels, I have found the fundamental issues customers face in today's turbulent energy world for buildings and facilities are universal. However, depending where you are, the relative importance may be dramatically different. The seven issues are the desire/need to:

- Use less energy—consume the least amount of energy while still performing the core mission;
- Reduce the cost of energy—buy energy at the lowest available unit cost;
- Stabilize energy costs—stabilize operating expense predictability;
- Reduce power outages—ensure energy quality and security;
- Infrastructural renewal—replace aging building/facility systems;
- Capital fund preservation—use capital funds for core business activities not for energy conservation and optimization; and
- Environmental responsibility—consume natural resources and managing waste production in an environmentally friendly way.

It takes a wide range of subject-matter-expertise, the previously described eight critical ESCO competencies, plus local, national, and regional (i.e., European Union), knowledge of business practices in many key areas. These areas include:

- Private and public sector capital project procurement standards
- Laws and regulations
- Tax and accounting codes
- Engineering and construction standards
- Human resources, employment standards and laws
- Cultural diversity and language diversity
- Local energy utility issues
- Plus many more

None of these are technology-based, rather all are intellectual property, which means people—and people have huge implications in the development of a new geographical market.

CHALLENGES

Most business people understand the challenges of trying to sell a product into a new geographical market. Is there a current market for our products, or will we be creating a new market? Can I create a brand identity? Does our current product meet local needs, or do we need to modify or create new products? Do we import from our current manufacturing sources, or do we establish new local manufacturing capacity? Can I protect my products' unique characteristics through patents or other legal mechanisms? Do we establish local partners to leverage existing distribution channels, or do we create new locally owned distribution channels? How will I provide post-purchase warranty and support? The questions go on and on.

When applying some of these basic questions to the ESCO business, some significant differences are instantly apparent.

We create a market through the disruption of established procurement methodologies of capital projects, because we sell and deliver our *"products and services"* directly to end-use customers.

Our "product" is uniquely created for each customer, and as such has to be "manufactured" locally.

We provide or help create the funding mechanism to purchase our "products."

We provide ongoing "warranty" (guaranteed performance) and additional services for the life of the project, in most cases for 7 to 15 years.

Our intellectual property and subject-matter-expertise is our product, and it will be difficult to protect.

Every ESCO that is looking to globalize their business is going to have to ask some very basic questions: Which and how many countries? Which vertical markets? What range of technologies? Funding and sustainability? Management and subject-matter-expertise acquisition? Buy, build, or partner, and the list goes on.

SUMMARY

All ESCOs have the ability to export best practices in customer knowledge, engineering, project management, and financial engineering to a new geography. If they are wise enough to add cultural sensitivity and receptivity to new ideas, they will have a good chance to succeed—if they can over come the local resource acquisition needs.

Unfortunately, many local and regional ESCO's will stumble in their efforts to develop a global ESCO business because of their mistaken belief in the universality of standard contracts, contract laws, procurement laws, procurement methodologies, decision-making criteria, and that funding mechanisms are in place.

Many will also stumble because they have "know it all attitudes," cultural biases, or closed-mindedness to different ways of doing business.

The difficulty of exporting the ESCO business to a new geography cannot be understated. It would appear that multi-national companies currently delivering products and services in a large number of geographic areas have an intrinsic and significant advantage. They already have an established business infrastructure in place. They have developed business, financial, legal, vendor and supplier, and political relationships already. They also have the financial strength to support the effort over the long haul. Whether they can evolve their local organizations into an ESCO, and have the patience to do so, will only be known over time.

Appendix A

Cost-effectiveness and Cost of Delay Applications

The material presented in Appendix A offers some practical applications in modifying the simple payback formula to accommodate O&M costs or variations in energy prices. Also offered below is the application of the Cost of Delay formula and its relationship to the Simplified Cash Flow financing formula.

1. **Changes in O&M costs**. The new equipment might require more, or less, O&M work. To the extent the O&M work can be quantified, the Adjusted Payback Period (APP) formula would then read as follows:

ADJUSTED PAYBACK FORMULA, O&M

$$APP_{O\&M} = \frac{I}{ES_n + M_n}$$

where,

$APP_{O\&M}$ = Payback period adjusted for O&M
I = Initial investment for period of analysis
ES_n = Energy savings for analysis period
M_n = Differential operations and maintenance costs for analysis period
n = Period of analysis

2. **Changes in energy costs**. Since the payback period is a function of energy costs, this approach can more accurately reflect the impact of

unstable energy prices. The difficulty comes in predicting future energy costs. Therefore, it is recommended that projections on energy prices be limited to three years and used only for internal discussion, unless the organization has a longer-term contract.

ADJUSTED PAYBACK FORMULA, ENERGY COST

$$APP_e = \frac{I}{X(E_{n1} + E_{n2} + E_{n3})}$$

where,

APP_e = Payback period adjusted for projected energy costs
I = Initial investment
E_{n1} = Projected 1st year energy costs
E_{n2} = Projected 2nd year energy costs
E_{n3} = Projected 3rd year energy costs

As long as utility prices are volatile, it is worth restating the inherent dangers in publicly justifying certain energy efficiency measures predicated on energy price increases (unless you are quoting data supplied by the utility or from a signed contract). If prices don't increase as predicted; or, worse yet, fall, you end up looking pretty bad. Even worse, the justification may become the story. The headline resulting from a university's board of trustees meeting may state, "20% Energy Price Increase Next Year: University Says." And the planned energy efficiency work becomes lost in the hullabaloo.

3. **Cost of Delay application**.

Using an annual simplified cash flow (SCF) formula without any adjustment for inflation and with the purchase/installation costs prorated over five years, the calculations would be:

SCF = (E + O&M savings for year) – (I [prorated])
= ($400,000) – ($1,400,000/5)
= $400,000 – $280,000
= + $120,000

The $400,000 savings would reduce the existing $1.6 M to $1.2 M. In this case, the potentially lower budget is used as a reference point.

CoD	=	minus (E + O&M for the year) + (I prorated)
CoD	=	The new lower budget minus the existing budget ($1.2 – $1.6 = – $400,000) Plus the prorated investment ($1,400,000/5 = $280,000)
	=	- $400,000 + $280,000
	=	- $120,000

In essence, this rather awkward formula looks at what could have been saved less the cost of the work required to achieve those savings over a specified period. To put it more simply, potential savings equal your losses if you do nothing.

The formula does not look beyond the years of prorated investment. In the above example, the CoD becomes $400,000 (exclusive of any O&M costs, depreciation or net present value) in the sixth year and every year thereafter, for the life of the improvement.

COST OF DELAY: SAMPLE PROBLEM

Data:

Annual utility bill	=	$1,500,000
O&M on existing equipment	=	600
Projected savings	=	30% energy 30% O&M
Retrofit investment	=	$1,750,000
Prorated period	=	5 years

Problem:

If the project is put off for one year, what is the Cost of Delay?

First, calculate the projected savings

($1,500,000 X 30%) + ($600 X 30%) = $450,000 + $180

Then, determine what it would cost each year to achieve those savings $1,750,000/5 years, or $350,000

(The net LOSS should be – $100,180/yr) $450,180 – 350,000

Now, calculate CoD for 6 years (without considering depreciation, time value of money, or adjustments in O&M.) ($1.5 million x 30%, or $450,000)

Figure 2-2. Cost of Delay Analysis

Using data supplied through energy audits, you can use the same Cost of Delay calculation procedures to determine what you may actually be losing every day, month, or year that your organization fails to act.

After the commitment to secure private sector financing has been made, delays are often incurred because outside support, particularly attorneys, are not familiar with the performance contracting process. These delays can be reduced if your attorney and engineer are involved early in the process.

Appendix B

Examples of Evaluation Criteria and Evaluation Procedures

CRITERION: ADEQUACY OF FINANCIAL ARRANGEMENT AND THE NET PRESENT VALUE OF COST SAVINGS TO THE DISTRICT

Factors To Be Considered:

a) Soundness of the estimate of the cost of services and equipment for the work proposed.

b) The degree to which the cost savings are guaranteed and the form of the guarantee.

c) Degree to which cost savings are based on measurable quantities, projections of baseline values or estimated quantities.

d) Use of energy price escalation; discount rates.

e) Terms of the sample contract provide optimum benefit to the organization (length, return on investment, payment schedule, share of savings, etc.)

f) Ability to finance project.

Score:

10 = Firm (or investor partner, joint venture partner, or subcontractor) has indicated that extensive capital and cash flow resources are avail-

able; terms of the sample contract are extremely favorable; the firm proposes the highest net present value of cost savings substantiated by documented case studies of similar results; the savings are fully guaranteed; and the organization risks no financial exposure.

8 = Financial resources are more than adequate; terms of the sample contract are favorable; the firm proposes a reasonable net present value of cost savings and they are based on quantitative measurements, reasonably solid estimates, or projections of baseline values and a history of delivering such value to its customer; the savings are mostly guaranteed and organizational risks are minimal.

6 = Financial resources appear adequate: terms of the sample agreement are adequate; the proposal offers strong cost savings with good references; the savings are mostly guaranteed or are reasonably sure of being achieved; and the organizational risks are limited.

4 = Financial arrangement and resources appear marginal; net present value of cost savings is low; and the organizational risks are of some concern.

2 = Financial arrangement and resources appear inadequate.

0 = There appear to be no financial resources for the firm; no cost savings are given in the proposal; or they are not likely to be achieved.

The following criteria are offered without scoring with the suggestion that the evaluation committee develop its own scoring to affirm that the process is well understood by all committee members.

CRITERION: TECHNICAL PERFORMANCE

Factors To Be Considered:

a) Comprehensiveness of approach.

b) Adequacy of the equipment to provide the services proposed and its integration into existing system(s).

c) Adequacy of the proposed operation and maintenance concept, including training of facility personnel.

d) Degree to which the proposed system meets all work environment requirements.

CRITERION: MANAGEMENT, SCHEDULE AND QUALITY ASSURANCE

Factors To Be Considered:

a) ESCO has been established as an entity long enough to assure it can deliver promised services and back them up. (Experience of *individuals* in a new ESCO are not sufficient—project management issues can still arise—and further assurances must be obtained.);

b) The organizational structure is clear and well-defined; the lines of communication are direct; and the management appears to be well informed and responsive.

c) The proposed schedule appears to be reasonable without being either too tight or dilatory; meshing of parallel and sequential activities is well orchestrated.

d) The proposed management structure and quality assurance program can identify problems promptly, and take effective remedial action.

e) The proposal clearly defines the organization's supplied resources; the proposed resources are reasonable for the project.

Appendix C
Sample Planning Agreement

____

Customer Name

____

Address or Location of Premises

____

City State Zip

______________**[ESCO] and the customer named above agree as follows:**

1. Energy and Operational Assessment

ESCO agrees to undertake a detailed evaluation study of the CUSTOMER'S premises identified above to assess energy consumption and operational characteristics of the premises and to identify the energy efficiency measures, procedures and other energy-related services that could be provided by ESCO in order to reduce the CUSTOMER'S energy consumption and operating costs on the premises.

2. Objectives

CUSTOMER agrees to provide its complete cooperation in the conduct and completion of the study.

ESCO will provide to the CUSTOMER a written report within 60 days of the effective date of this Agreement. The report will meet the following objectives:

(a) a list of specific energy efficiency measures that ESCO proposes to install with estimated acquisition and installation costs;

(b) such measures will not negatively impact on the work environment or any industrial process;

(c) a description of the operating and maintenance procedures that ESCO believes can reduce energy consumption and operating costs at the premises; and

(d) an estimate of the energy and operations costs that will be saved by the services, equipment and procedures recommended in the report.

3. Records and Data

CUSTOMER will furnish to ESCO upon its request, accurate and complete data concerning energy usage and operational expenditures for the premises needed for the energy and operational assessment, including the following data for the most recent two years from the effective date of this agreement:

- actual utility bills supplied by the utility;
- other relevant utility records;
- descriptions of all energy-consuming or energy-saving equipment used on the Premises;
- descriptions of any recent changes in the building structure or its energy consuming systems, including heating, cooling, lighting;
- occupancy and usage information;
- descriptions of energy management and other relevant operational or maintenance procedures utilized on the premises;
- summary of expenditures for outsourced maintenance, repairs or replacement on the premises;
- copies of representative current tenant leases, if any; and

- prior energy audits or studies of the premises, if any.

4. Preparation of Energy Services Agreement (ESA)

Within 30 days after the submission to ESCO of the report described under paragraph 1 of this agreement, ESCO will prepare and submit to the CUSTOMER an ESA to implement the energy efficiency measures, procedures, and services identified in the report that could reduce the CUSTOMER'S energy consumption in the premises.

5. Payment Terms

CUSTOMER agrees to pay to ESCO a sum specified in the audit report not to exceed $________, within 60 days after the delivery to the CUSTOMER of the report described under paragraph 1 of this Agreement. However, CUSTOMER will have no obligation to pay this amount if:

(a) ESCO and the CUSTOMER enter into an ESA within 60 days after the delivery to the CUSTOMER of the report described under paragraph 1 of this Performance Contracting Planning Agreement;

(b) An independent engineer, approved by the ESCO prior to the audit, deems that the audit has not adequately assessed the CUSTOMER'S energy efficiency opportunities; and/or met the objectives described in Section 2 of this agreement.

(c) A majority of the recommended energy efficiency measures cannot be implemented without a negative impact on the CUSTOMER'S processes and/or comfort conditions.

6. Standards of Practice

The detail and quality of the ESCO's investment grade audit and planning agreement as previously submitted are representative of its work shall serve as a standards of practice for all work under this project.

7. Indemnity

ESCO and the CUSTOMER agree that ESCO shall be responsible only for such injury, loss, or damage caused by the intentional misconduct or the negligent act or omission of ESCO or its agents in performing the work described herein. ESCO and the CUSTOMER agree to indemnify and to hold each other, including their officers, agents, directors, and employees, harmless from all claims, demands, or suits of any kind, including all legal costs and attorney's fees, resulting from the intentional misconduct of their employees or any negligent act or omission by their employees or agents.

8. Arbitration

If a dispute arises under this agreement, the parties shall promptly attempt in good faith to resolve the dispute by negotiation. All disputes not resolved by negotiation shall be resolved in accordance with the Commercial Rules of the American Arbitration Association in effect at the time, except as modified herein. All disputes shall be decided by a single arbitrator. The arbitrator shall issue a scheduling order that shall not be modified except by the mutual agreement of the parties. Judgment may be entered upon the award in the highest state or federal court having jurisdiction over the matter. The prevailing party shall recover all costs, including attorney's fees, incurred as a result of the dispute.

9. Assignment

This agreement cannot be assigned by either party without prior written consent of the other party. This agreement is the entire Agreement between ESCO and the CUSTOMER and supersedes any prior oral understandings, written agreements, proposals, or other communications between ESCO and the CUSTOMER.

10. Miscellaneous Provisions

Any change or modification to this agreement will not be effective unless made in writing. This written instrument must specifically indicate that it is an amendment, change, or modification to the agreement.

ESCO	**CUSTOMER**
By	By
______________________	______________________
Signature	Signature
______________________	______________________
Title	Title
______________________	______________________
Date	Date

Glossary of Terms

BASELINE—Annually adjusted baseyear

BASEYEAR (historical consumption)—A recent year's, or average of __ years' energy consumption and operating conditions affecting consumption; used as a reference base to compute savings attributable to the energy efficiency measures. Some professionals use baseline for baseyear and adjusted baseline to designate annual baseline.

CHAUFFAGE—Combined supply and demand efficiency services on a guaranteed basis; e.g., conditioned space at a cost per square foot (or meter) or energy use per unit(s) of production

COMMISSIONING—The performance verification, fine tuning, maintenance protocols, etc., associated with the new construction and/or major renovation of a building.

COST OF DELAY—The net loss from energy efficiency work not performed.

DEMAND SIDE MANAGEMENT—The efficient management of energy use on the demand (customer) side of the meter.

ENERGY PERFORMANCE CONTRACTING—Synonymous with the term performance contracting; more widely used in Europe

ENERGY SERVICE COMPANY (ESCO)—A firm which provides energy management services including an engineering evaluation of the building; installation of energy-saving equipment and modifications; maintenance procedures; and financing arrangements at an agreed upon comfort level for a fee usually guaranteed not to exceed the cost of the project.

ENERGY SERVICE PROVIDER—An energy service company or registered professionals, such as architectural and engineering firms, that provide the expertise, services, equipment, modifications, and financing without performance contracting guarantees.

GUARANTEED SAVINGS—Agreement, or contract clause, whereby a firm will guarantee that a piece of equipment, or a package of energy efficiency measures, will achieve a minimum amount of savings over a contract period. Also refers to a performance contracting model whereby a firm will arrange customer financing and will refund the difference between the actual and guaranteed savings, typically the debt service amount.

HURDLE RATE—The point at which it becomes financially beneficial to the customer to use outside funds; usually based on present rate of interest earned on internal funds and related conditions.

INSTALLMENT/PURCHASE—An arrangement whereby the purchaser makes payments at regular intervals until the cost of the equipment has been satisfied. Unlike lease/purchase, ownership is predetermined and the contract MAY exceed a fiscal year without automatic renewal provisions.

INTERNAL FINANCING—Financing with money available or retained by the organization without securing outside revenues.

INTERNAL RATE OF RETURN—Discounts an investment's expected net cash flow to a net present value of zero.

LEASE—An agreement to make regular payments to a lessor over a set period of time in exchange for the use of a building or equipment.

LEASE PURCHASE (closed-end lease)—An arrangement whereby a lessee commits to making payments for the use of buildings or equipment for a set period of time. The lessee has the right to buy the property for a price agreed upon in advance, frequently a nominal figure.

LESSEE—The user of a leased asset who pays the lessor for the usage right.

LESSOR—The owner of a leased asset.

LOAD MANAGEMENT—Actions taken to reduce peak demand through load shedding and load shifting.

LOAD PROFILE—The average electric demand divided by the peak demand.

MINIMUM ACCEPTABLE RATE OF RETURN—The lowest rate of return that an investment can be expected to earn and still be acceptable; same as the investment's cost of capital.

NET PRESENT VALUE (NPV) (of energy savings)—The value in today's currency (year zero) of future energy savings less all project contracting, financing, and operating costs. This measure takes into account the "time value of money.

OPEN-ENDED LEASE—A lease where there is no fixed price purchase option at the conclusion of the lease period. (Lease may indicate "purchase at fair market value," but no specific amount is stated.)

OPERATING LEASE—A lease that does not meet capital lease criteria and is cancelable by the lessee at any time upon due notice to the lessor. Also refers to a short-term lease that is cancelable by the lessor or lessee upon due notice to the other party.

PAYBACK PERIOD—The amount of time required for an asset to generate enough net positive cash flow to cover the initial outlay for that asset.

PERFORMANCE CONTRACT—A contract with payment based on performance; usually guarantees that project costs will not exceed projected savings.

POSITIVE CASH FLOW LEASE—A closed-end lease (lease/purchase) in which payments made to the lessor are kept below the level of savings derived from the leased equipment (for each lease payment, a specified time period, or the total lease period.)

POWER MARKETING—Managing products and services associated with electrical power; their delivery for fixed or variable prices .

PRESENT VALUE—The value of money at a given date (current value) that will be paid or received in future periods.

PRIME RATE—Interest rate charged by banks on short-term loans to large low-risk businesses.

PROJECT SAVINGS (in a savings based financing agreement)—Refers to the expected annual financial value of the reduced energy consumption due to implementing energy efficiency measures.

RATE OF RETURN—The interest rate earned on an investment; may be the actual rate or expected rate.

RISK SHEDDING—Procedures used to assign risks to another party.

SALVAGE VALUE—(residual value)—The money that remains in an asset after it has been held/used for a period of time.

SAVINGS-BASED AGREEMENT (OR CONTRACT)—Arrangements by which energy service companies agree to provide services and energy efficiency improvements in a client's building or industrial process with the repayment to come from savings generated by the improvements.

SAVINGS-BASED FORMULA—The formula (calculation of savings procedure) specified in the contract which is used to determine savings. Usually involves four steps: (1) determine actual historical usage and contributing operating conditions to form a baseyear; (2) adjust baseyear (usually annually) to actual usage for variations (temperature, occupancy, etc.) to form a baseline; (3) subtract actual usage from adjusted baseline consumption; and (4) calculate savings by multiplying the units of energy saved by the current cost per current cost per unit. Calculations for electrical demand savings are considered part of the formula but computed separately.

SELF-FUNDING—A project or program financed through savings achieved by project actions or measures taken.

SHARED SAVINGS—A financial model for performance contracting where a predetermined split of the energy *cost* savings is shared between customer and ESCO; usually the ESCO carries the financing and the credit risk

SENSITIVITY ANALYSIS—Analysis of the effect on a project's cash flow (or profitability) of possible changes in factors which affect the project; e.g., level of predicted savings, energy price escalation, etc.

STRANDED BENEFITS— Societal, environmental and economic benefits frequently provided by utility voluntarily or by order, which are or will be lost when utilities are restructured and will operate in a competitive environment.

STRANDED COSTS—Debts incurred to construct generating capacity which have not been fully retired.

SUPPLY SIDE MANAGEMENT—The efficient management of electrical generation, transmission and distribution to the customer's meter.

Index

A

AcquiSuite 246, 251, 252
active server pages (ASP) 248
adjusted baseline 116
adjusted payback formula
  energy cost 306
  O&M 305
Ahmedabad 284
air changes per hour 42
air handlers 258
Akuja
Anil 148
all-purpose ESCO 221
Allen, J. x, 215
An ESCO's Guide to Measurement and Verification 1
arbitration 120
ASHRAE-14 69
ASHRAE-62 203, 206
ASHRAE Guideline 1-1996 191
Asian Development Bank (ADB) 28, 275, 280
asset-based lending 294
asset monetization 226
attorneys 123, 129, 141
audiences
  target 149
audits 48
avoided utility costs 14

B

"buy" signals 184
bankable projects 89, 187
banker 96
baseline 80, 111, 252
  adjusted 116
  development 80
baseyear 83, 111, 116, 122
baseyear/baseline 188
Bechtel 281
Beijing 284
beltway bandits 50
Bertoldi, Paolo 18
BGA, Inc. 215, 231
bid/spec 46
bidders' conference 48
Birr, David 143
board 149, 150, 155
bond ratings 169
Boston University 28
brand identity 303
Brazil ix, 280, 285
Brown, James W. 29, 186
building code 79
Building Manager Online 250
building occupants 146, 155
Bulovka 290
  Hospital 289
bundled solution 225

burn money 8
Burroughs, H.E. 201
business councils 278
business customs 271
buy-down 99

C

C&I 221
capacity building 295
capex 16
capitalized interest 103
capital lease 98
cash-poor customers 217
cash flow 23
Central and Eastern Europe 280
Central and South America 280
certificates of participation (COP) 102
Certified Measurement and Verification Professional (CMVP) 68
CGI 249
chauffage 22, 26, 223
chilled water bypass valve 258
chiller plant 257, 258, 259, 262
China 275, 280, 283, 291
Chinese local banks 292
CMMS 95, 193
CO_2 42, 206, 210, 258
comfort standards 117
commencement date 119
commercial and industrial 218
commercial business 16
commissioning 82, 83, 190
    process 190
commodity 229
common gateway interface 247
communications 143
    strategies 156
    techniques 146
Compagnie Generale de Chauff 21
compensation 118
computer-based maintenance management software 94
computer-based maintenance management system 193
construction manager 128, 189
content outline 52
continuous commissioning 191
contracts 106
convertibility of currency 278
cost-effectiveness 31, 305
cost/benefit analysis 9
cost avoidance 31, 38, 39
    graph 39
cost effectiveness 32
cost of acquisition 185
cost of delay 31, 35, 37, 38, 169, 305
    application 306
cost of issuance 104
country analysis 277
credit 164
    review 297
    risk 170
creditworthiness 91, 169
criteria 43, 45, 52, 53, 54, 309
critical risk 180
culture 284
customer pre-qualification 180
cutting edge technology 224
Czech Republic 289

D

DAS 246
data 252
    acquisition 243
    acquisition server 246, 251, 254
    analysis 259

collection process 244
Dawson, Roger 123, 124, 126
deal killer 289
deal stoppers 184
debt ceiling 169
debt service obligation 126
decision matrix 56
decision schedule 184
defaults 115
and remedies 120
deferred maintenance 209
design-bid-build 235, 236
design build 229, 236
process 237
DG JRC 18
Dialight 234
LLC 215
dilution delusion 204
direct digital controls 69
direct purchase 166, 229
discounted 172
discount rates 170, 309
risks 172
Dixon, Bob x, 265
Douglas, Jeannie C. Weisman ix
Dreessen, Tom x, 264
DSM 284
due diligence 93

E

Eastern Europe 281, 289
economic climates 271
EEPs 293, 294, 295, 298
effective communications 145
elements of 146
effective risk strategies 197
efficiency 143
Efficiency Benefits Fund 299
Efficiency Valuation Organization (EVO) 296
EIS 243, 244, 245, 247
implementation 249
server 246
service 250
electric submeter 251, 254
emergency response 176
emission 3
employees 155
EMS 246
for data collection 244
energy 15
audits 308
conservation 143, 201
dependence 15
efficiency 201
projects (EEPs) 291, 292
sell 5
management 131
companies 291
plan 138
manager 128, 130, 133, 137, 138, 139, 143, 144, 148, 194
manager's job 138
prices 172
security 140
services 19
agreement (ESA) 21, 109, 314
agreement components 110
Energy Information System (EIS) 243, 244
Energy Management System (EMS) 244
Energy Masters 263
Energy Systems Group 1
engineer 141
Enron 224, 281
environment 9, 147, 301
environmental mandates 9

EPS Capital Corporation 264
equipment 188
  ownership 113
  selection and installation 114
Ernest Orlando Lawrence Berkley (LBL) National Laboratory 230
ESCO fee 198
ESCO management strategies 129
ESCO risks 179
Ethernet 245
European Bank 28
European Bank for Reconstruction and Development (EBRD) 275, 280
European Commission 18
European Union 302
evaluating proposals 53
evaluation procedures 44, 309
evaluation process 45, 50, 53, 56
evaluation criteria 309
excess savings 195
extended warranty 238
external publics 152

F

facility/technical factors 182
facility blind 8
facility control 174
facility managers 6
factors 309
FASER™ 84
Federal Energy Management Program 35
federal government 27
FERC 224
financial/economic factors 181
financial information 91
financial institutions
  local 294
financial risks 166
  framework 167, 171
financial structure 195
financing 100
  mechanisms 100
firm's qualifications 50
fixed costs 219
floor price 23
flow meter 252
formulas 116
French model 24
fresh, natural air 202

G

general contractor 189
general obligation bonds 100
Germany 284
Gibson, Michael x, 215
globalization 275, 301
Global Environment Fund 291
global market 275
Green, David C. x, 215
Green Management Services 215
guaranteed performance 16, 303
guaranteed savings 18, 23, 163, 164, 166
guarantees 118, 121, 179

H

Hanneman Hospital 17
Hansen, Jim x, 264
Healthy Buildings Institute 209
hidden project costs 172
Holmes, Oliver Wendell 44
Honeywell 168, 209
Hong Kong 219
Hopper, Nicole 242
hospital 16
human element 185

human factors 187
humidity control 259
hurdle rates 5, 219

I

IAQ 159, 201, 206, 208, 211
  energy efficiency relationship 208
  problems 210
IEEFP 296, 298
IGA 92, 108, 132, 186, 187, 188, 196
in-country partners 286, 288
indemnification 111, 115, 120, 314
India 275, 280, 281
indoor air quality 159, 201
industrial 218
  opportunity 218
industry 16, 218
infiltration 211
information technology 243
integrated solutions 22, 26, 223
Inter-American Development Bank 28, 280
international 267, 268, 269
  financing 291
  market 267, 276
  opportunities 275
International Bank of Reconstruction and Development 291
International Energy Efficiency Project Financing Protocol (IEEFP) 296
International Finance Corporation 291
International Performance Measurement & Verification Protocol (IPMVP) 69
Internet 247
interviews 49
Intranet 247
intrusion/interruption 169
investment grade audit (IGA) 92, 107, 186, 185, 188
investment grade energy audit 21
IPMVP, Inc. 69, 70, 71, 296
IRRs 5, 219
IT 243, 245

J

JavaScript 248
Java Applets 248
Johnson Controls 168

K

Kannberg, Dr. Landis 140
key competencies 301
key contract considerations 113
Kiona 264
Kiona International 264

L

LAN 245, 246
laws 271
LCC 34, 35
Leadership in Energy and Engineering Design (LEED) 68
lean and clean management 147
lease-based financing 101
leases 98
lease revenue bonds 103
LEED 68
legal issues 279
Lewis, Jim x, 215, 250
Li, Mr. 286
life-cycle costing 34, 187
load management 19
local financial institutions (LFIs) 293

proposals 62
preparation 62
public benefits charge 299
purpose and scope 52

Q

quality assurance 311
quick fix 26

R

re-circulated air 205
recommissioning 191
reconstruction and development 28
recycling 226
Reedy Creek Energy Services 215
relative humidity 204
repatriation 278
requests for qualifications format 51
request for proposal 41
request for qualifications (RFQs) 41, 46, 107
reserve fund 103
results 43
retrocommissioning 191
return on assets 227
return on investment 97, 170
revenue bonds 101
Rezessy, Sylvia 18
RFPs 42, 43, 46, 47, 49, 107
team 43
RFQ 49
RFQ/RFP 152
risk 20, 157, 159, 161, 163, 176, 179, 189
analysis framework 161
control 268
cushion 195
factors 166, 174
management 157, 158
shedding 176
vulnerability 179
risk/benefit ratio 158
ROIs 5, 219, 252, 271
Romm, Joseph J. 27, 147
Roosa, Stephen A. ix, 1, 95
Royal Dutch Shell 17
Russia 213, 282

S

San Francisco
City and County of 163
savings-based project financing 294
savings calculations formulas 116
savings persistence 172
schedules 109, 121
Schlesinger, James 213
school 16
scoping audit 91
scores 54
secrets of power negotiating 123
self-funded 3
Shakespeare 213
shared savings 17, 18, 22, 24, 99, 163, 164, 166, 168, 170
short circuited 205
sick building syndrome 203
Siemens 168, 265
simple payback 33
period (SPP) 33
simplified cash flow 33, 34
single purpose entity (SPE) 110
single purpose vehicles (SPV) 99
Smith, Don 263
sources of IAQ problems 209
special assessment 100

special purpose entities (SPE) 24, 99
special purpose vehicle (SPV) 24
specifications 46, 125
speculate 156
Square D 247
staff 153
standards of practice 314
state agency 16
State Economic and Trade Commission 286
State Energy Office 36
state government 27
stationary fuel cells 26
statute 169
stolen power 282
structuring the deal 21
subcontractors 189
supply acquisition 222, 224
surety 88
sustainability 303
sustainable development 212

T

tax-exempt financing 170
tax-exempt organization 97, 110
technical performance 310
technical risk 164
  framework 165
temperature sensors 252
termination 114, 120
test and balance 261
test audits 47
third-party financing 97, 271
tigers 286
tight building syndrome 206
Time Energy 17
time value of money 172
Todd, Brian x, 215
top management 7
total retrofit market 270
TPF 97
traditional energy audit 185
transitional economies 4, 24
Tula 284
two-stage solicitation 50

U

U.K. 281
U.S. Agency for International Development (AID) 27, 274, 287
U.S. Chamber of Commerce 274, 278
U.S. Commerce Department 278
U.S. Green Building Council 84
Ukraine 213, 222
United Nations Development Program ix
USAID 291
US Agency for International Development (USAID) ix, 275
US Department of Commerce 35
US Department of Energy 35
US Export-Import Bank 280

V

value chain 26
Vance, Christine 163
variable speed drives 258
VBScript 248
vendor 229, 233, 234, 237, 241
  financing 163, 164, 166, 170
ventilation 202, 203, 208
  for acceptable indoor air quality 203
  mitigation 206
verification 65
Veris 247

local government 27
local partner 272, 303
Lombardo Associates/ERI Services 28
low bid 173

M

M&V 65, 193, 292
  options 65
  option application matrix 78, 79
M&V plan 77, 78, 79, 83
M&V protocols 69
M&V techniques 67
maintenance 94
malfunctions 113
management 6, 7, 8, 10
  strategies for the owner 134
managing indoor air quality 201
managing risks 195
margin 12
market-based incentives 295, 298
marketing channel 229
marketing systems 268
market assessment 288
market channel 235
market conditions 295
market resistance 202
master leases 102
material change 120
MDBs 291, 293
Means, R.S. 173
measure-specific risks 179
measurement 65
measurement and verification (M&V) 65, 250
  guideline for federal energy projects 69
  plan 75, 77
  protocol 193
Meckler, Milton 205
Mello-Roos bonds 100
METRIX™ 84
mitigated 179
mitigating/managing 158
mitigating strategies 158, 161, 174
Modbus RTU 247
Mongolia ix
Moscow 284
Moscow Power Engineering Institute 282
multi-lateral development banks (MDBs) 280, 291
multi-national companies 304
municipal leases 102

N

National Association of Energy Services Companies 265, 274
National Energy Conservation Policy Act 27
National Institute of Occupational Safety and Health (NIOSH) 206, 209
natural air 208
negotiations 122
net present value 187
new geographical market 303
NIOSH 206, 209
North America 25

O

O&M 7, 182, 192, 210
  costs 305
  functions 192
  measures 7
  personnel 6
  staff 155, 176, 187
  tasks 192

Obvius 215, 246
occupant behavior 186
OEM 234
off balance sheet 170
on-site treatment facility 226
OPEC 213
open book 46, 173
operating costs 10, 11
operating lease 98
operating system (OS) 246
operational savings 20
operations and maintenance 8, 154
  practices 192
  training 174
Option A 70
Option B 71
Option C 71
Option D 71
org. chart 137
outside air 205
outside audiences 152
outsourcing 19, 227
ownership 119
owner risks 161

P

paid from savings 21
partnership 128
paybacks 185
payback criteria 172
people factors 91, 180, 182
perceived risks 195
performance bonds 91
performance risks 164
performance standards 122
PERL 248
Philippines 281
photo opportunities 153
picking up the pace 163
planning agreement 21, 60, 107, 108, 312
PM/EM relationship 133
Poland 282
policy 135
  statements 136
pollutants 3
pollution emissions 5
positive cash flow 195
post-implementation report 82
post contract savings 168
power marketing 19
Power Measurement Ltd. 247
pre-construction assessment 75
  M&V 76
pre-proposal conference 48
pre-qualification criteria 91, 181
pre-qualifying customer 90
predictive consistency 63, 93
present value 172
price of caution 196
privatization 226
procedural/administrative risks 179
procedural risks 173
  framework 175
products and services 301, 302
product warranty vs. PC contract term 234
project assessment 141
project delivery 231
project development 183
project evaluation 130
project financing 271
project management 93, 174
project manager 95, 128, 130, 131, 132, 133, 143, 148, 194
project margin pressure 232
project timing 233

Viron 263

W

Walt Disney World 215
warranty 121, 166, 303
Warsaw 284
wastewater 226
  systems 28
water conservation measures 76
water management 19
water resource management 28
water set point 258
web 243
  browser 244
  publishing 244, 247
    process 245
weighting 45, 52
WISHCOs 91
World Bank ix, 28, 275, 278, 280, 291, 292

X

XYZ Shoe Manufacturers 12